Lecture Notes in Computer Science 9499

Commenced Publication in 1973
Founding and Former Series Editors:
Gerhard Goos, Juris Hartmanis, and Jan van Leeuwen

More information about this series at http://www.springer.com/series/7407

Laura Sanità · Martin Skutella (Eds.)

Approximation and Online Algorithms

13th International Workshop, WAOA 2015
Patras, Greece, September 17–18, 2015
Revised Selected Papers

Springer

Editors
Laura Sanità
University of Waterloo
Waterloo
Canada

Martin Skutella
Technische Universität Berlin
Berlin
Germany

ISSN 0302-9743   ISSN 1611-3349   (electronic)
Lecture Notes in Computer Science
ISBN 978-3-319-28683-9   ISBN 978-3-319-28684-6   (eBook)
DOI 10.1007/978-3-319-28684-6

Library of Congress Control Number: 2015958903

LNCS Sublibrary: SL1 – Theoretical Computer Science and General Issues

Preface

The 13th Workshop on Approximation and Online Algorithms (WAOA 2015) focused on the design and analysis of approximation and online algorithms. These algorithms have become a fundamental tool in several fields and in many applications that cope with computationally hard problems and problems in which the input is gradually disclosed over time.

WAOA 2015 took place in Patras (Greece) during September 17–18, 2015, and it was part of ALGO 2015, which also hosted ESA, ALGOCLOUD, ALGOSENSORS, ATMOS, IPEC, and MASSIVE. The previous WAOA workshops were held in Budapest (2003), Rome (2004), Palma de Mallorca (2005), Zurich (2006), Eilat (2007), Karlsruhe (2008), Copenhagen (2009), Liverpool (2010), Saarbrücken (2011), Ljubljana (2012), Sophia Antipolis (2013), and Wraclaw (2014). The proceedings of all these previous WAOA workshops have been published as LNCS volumes.

Topics of interest for WAOA 2015 were: algorithmic game theory, algorithmic trading, coloring and partitioning, competitive analysis, computational advertising, computational finance, cuts and connectivity, geometric problems, graph algorithms, inapproximability, mechanism design, natural algorithms, network design, packing and covering, paradigms for the design and analysis of approximation and online algorithms, parameterized complexity, scheduling problems, and real-world applications. In response to the call for papers, we received 40 submissions, one of which was subsequently withdrawn. Each submission was reviewed by at least three referees, and mainly judged on originality, technical quality, and relevance to the topics of the conference. Based on the reviews, the Program Committee selected 17 papers. This volume contains final revised versions of these papers.

We would like to thank all the authors who submitted papers to WAOA 2015, and our plenary invited speaker Jochen Könemann for accepting our invitation. Furthermore, we are extremely grateful to the members of the Organizing Committee and its chair, Christos Zaroliagis, for doing a superb job with organizing ALGO 2015.

October 2015

Laura Sanità
Martin Skutella

Organization

Program Committee

Leah Epstein	University of Haifa, Israel
Samuel Fiorini	Université Libre de Bruxelles, Belgium
Zachary Friggstad	University of Alberta, Canada
Anupam Gupta	Carnegie Mellon University, USA
Lap Chi Lau	University of Waterloo, Canada
Nicole Megow	Technische Universität München, Germany
Marco Molinaro	TU Delft, The Netherlands
Britta Peis	RWTH Aachen University, Germany
Thomas Rothvoss	University of Washington, Seattle, USA
Laura Sanità	University of Waterloo, Canada
Jiří Sgall	Charles University, Prague, Czech Republic
Martin Skutella	Technische Universität Berlin, Germany
José Soto	Universidad de Chile, Chile
Kavitha Telikepalli	Tata Institute of Fundamental Research, Mumbai, India
Andreas Wiese	Max Planck Institute for Informatics, Germany
David Williamson	Cornell University, USA
Gerhard Woeginger	TU Eindhoven, The Netherlands
Paul Wollan	University of Rome, Italy
Rico Zenklusen	ETH Zürich, Switzerland

Additional Reviewers

Abdi, Ahmad	Dósa, György
Adamaszek, Anna	Erlebach, Thomas
Adjiashvili, David	Fekete, Sándor
Al-Bawani, Kamal	Folwarczný, Lukáš
Antoniadis, Antonios	Galanis, Andreas
Azar, Yossi	Gottschalk, Corinna
Batra, Jatin	Gupta, Manoj
Boyar, Joan	Han, Xin
Bruhn, Henning	Heydrich, Sandy
Büsing, Christina	Imreh, Csanad
Cardinal, Jean	Jerrum, Mark
Chalermsook, Parinya	Klimm, Max
Chestnut, Steve	Kwok, Tsz Chiu
Chrobak, Marek	Künnemann, Marvin
De Keijzer, Bart	Lokshtanov, Daniel

Matuschke, Jannik
Mömke, Tobias
Niazadeh, Rad
Pruhs, Kirk
Pérez-Lantero, Pablo
Roy, Bodhayan
Schaudt, Oliver
Schwartz, Roy
Spoerhase, Joachim
Stee, Rob van

Sviridenko, Maxim
Tiedemann, Morten
Tönnis, Andreas
Uetz, Marc V.
Venkat, Rakesh
Verschae, José
Veselý, Pavel
Weil, Vera
Wierz, Andreas
Zych, Anna

Contents

Approximation Algorithms for k-Connected Graph Factors

Bodo Manthey[✉] and Marten Waanders

University of Twente, Enschede, The Netherlands
b.manthey@utwente.nl

Abstract. Finding low-cost spanning subgraphs with given degree and connectivity requirements is a fundamental problem in the area of network design. We consider the problem of finding d-regular spanning subgraphs (or d-factors) of minimum weight with connectivity requirements. For the case of k-edge-connectedness, we present approximation algorithms that achieve constant approximation ratios for all $d \geq 2 \cdot \lceil k/2 \rceil$. For the case of k-vertex-connectedness, we achieve constant approximation ratios for $d \geq 2k - 1$. Our algorithms also work for arbitrary degree sequences if the minimum degree is at least $2 \cdot \lceil k/2 \rceil$ (for k-edge-connectivity) or $2k - 1$ (for k-vertex-connectivity).

1 Introduction

Finding low-cost spanning subgraphs with given degree and connectivity requirements is a fundamental problem in the area of network design. The usual setting is that there are connectivity and degree requirements. Then the goal is to find a cheap subgraph that meets the connectivity requirements and the degree bounds. Beyond simple connectedness, higher connectivity, such as k-vertex-connectivity or k-edge-connectivity, has been considered in order to increase the reliability of the network. Most variants of such problems are NP-hard. Because of this, finding good approximation algorithms for such network design problems has been the topic of a significant amount of research [1, 4–10, 14, 16–20].

In this paper, we study the problem of finding low-cost spanning subgraphs with given degrees that meet connectivity requirements (they should be k-edge-connected or k-vertex-connected for a given k). Violation of the degree constraint is not allowed. While d-regular, spanning subgraphs of minimum weight can be found efficiently using Tutte's reduction to the matching problem [21,23], even asking for simple connectedness makes the problem NP-hard [2]. For instance, asking for a 2-regular, connected graph of minimum weight is the NP-hard traveling salesman problem (TSP) [11, ND22].

1.1 Problem Definitions and Preliminaries

Graphs and Connectivity. All graphs in this paper are undirected and simple. Let $G = (V, E)$ a graph. In the following, $n = |V|$ is the number of vertices.

L. Sanità and M. Skutella (Eds.): WAOA 2015, LNCS 9499, pp. 1–12, 2015.
DOI: 10.1007/978-3-319-28684-6_1

For a subset $X \subseteq V$ of the vertices, let $\text{cut}_G(X)$ be the number of edges in G with one endpoint in X and the other endpoint in $\overline{X} = V \setminus X$. For two disjoint sets $X, Y \subseteq V$ of vertices, let $\text{cut}_G(X, Y)$ be the number of edges in G with one endpoint in X and the other endpoint in Y.

Two vertices $u, v \in V$ are *locally k-edge-connected in* G if there are at least k edge-disjoint paths from u to v in G. Equivalently, u and v are locally k-edge-connected in G if $\text{cut}_G(X) \geq k$ for all $X \subseteq V$ with $u \in X$ and $v \notin X$. Local k-edge-connectedness is an equivalence relation as it is symmetric, reflexive, and transitive. A graph G is *k-edge-connected* if all pairs of vertices are locally k-edge-connected in G.

Let $X \subseteq V$. We call X a *k-edge-connected component* of G if the graph induced by X is k-edge-connected. We call X a *locally k-edge-connected component of* G if all $u, v \in X$ are locally k-edge-connected in G. Note that every k-edge-connected component of G is also a locally k-edge-connected component, but the reverse is not true.

A graph G is *k-vertex-connected*, if the graph induced by the vertices $V \setminus K$ is connected for all sets $K \subseteq V$ with $|K| \leq k - 1$. Equivalently, for any two non-adjacent vertices $u, v \in V$, there exist at least k vertex-disjoint paths connecting u to v in G.

For an overview of connectivity and algorithms for computing connectivity and connected components, we refer to two surveys [13,15].

For a vertex $v \in V$, let $N_G(v) = N(v) = \{u \in V \mid \{u, v\} \in E\}$ be the neighbors of v in G. The graph G is *d-regular* if $|N(v)| = d$ for all $v \in V$. A d-regular spanning subgraph of a graph is called a *d-factor*.

By abusing notation, we identify a set $X \subseteq V$ of vertices with the subgraph induced by X. Similarly, if the set V of vertices is clear from the context, we identify a set F of edges with the graph (V, F).

Problem Definitions. Let $G = (V, E)$ be an undirected, complete graph with non-negative edge weights w. The edge weights are assumed to satisfy the triangle inequality, i.e., $w(\{u, v\}) \leq w(\{u, x\}) + w(\{x, v\})$ for all distinct $u, v, x \in V$. For some set $F \subseteq E$ of edges, we denote by $w(F) = \sum_{e \in F} w(e)$ the sum of its edge weights. The weight of a subgraph is the weight of its edge set.

The problems considered in this paper are the following: as input, we are given G and w as above. Then **Min-dReg-kEdge** denotes the problem of finding a k-edge-connected d-factor of G of minimum weight. Similarly, **Min-dReg-kVertex** denotes the problem of finding a k-vertex-connected d-factor of G of minimum weight.

Some of these problems coincide:

- Min-dReg-1Edge and Min-dReg-1Vertex are identical for all d since 1-edge-connectedness and 1-vertex-connectedness are simple connectedness.
- For $k \in \{1, 2\}$, the problems Min-2Reg-kEdge and Min-2Reg-kVertex are identical to the traveling salesman problem (TSP) because of the following: 2-regular graphs consist solely of simple cycles. If they are connected, they are 2-vertex-connected and form Hamiltonian cycles.

- For even d and k, the problems Min-dReg-$(k-1)$Edge and Min-dReg-kEdge are identical. For even d, every d-factor can be decomposed into $d/2$ 2-factors. Thus, the size of every cut is even. Therefore, every d-regular $(k-1)$-edge-connected graph is automatically k-edge-connected for even k.
- For $k \in \{1, 2, 3\}$, the two problems Min-3Reg-kEdge and Min-3Reg-kVertex are identical since edge- and vertex-connectivity are equal in cubic graphs [24, Theorem 4.1.11].

We also consider the generalizations of the problems Min-dReg-kEdge and Min-dReg-kVertex to arbitrary degree sequences: for **Min-dGen-kEdge**, we are given as additional input a degree requirement $d_v \in \mathbb{N}$ for every vertex v. The parameter d is a lower bound for the degree requirements, i.e., we have $d_v \geq d$ for all vertices v. The goal is to compute a k-edge-connected spanning subgraph in which every vertex v is incident to exactly d_v vertices. **Min-dGen-kVertex** is analogously defined for k-vertex-connectivity. For the sake of readability, we restrict the presentation of our algorithms in Sects. 2 and 3 to Min-dReg-dEdge and Min-dReg-kVertex, respectively, and we state the generalized results for Min-dGen-kEdge and Min-dGen-kVertex only in Sect. 4.

We use the following notation: OptE^k denotes a k-edge-connected spanning subgraph of minimum weight. OptV^k denotes a k-vertex-connected spanning subgraph of minimum weight. For both, no degree requirements have to be satisfied. OptF_d denotes a (not necessarily connected) d-factor of minimum weight. OptEF_d^k and OptVF_d^k denote minimum-weight k-edge-connected and k-vertex-connected d-factors, respectively.

We have $w(\mathrm{OptF}_d) \leq w(\mathrm{OptEF}_d^k) \leq w(\mathrm{OptVF}_d^k)$ since every k-vertex-connected graph is also k-edge-connected. Both $w(\mathrm{OptEF}_d^k)$ and $w(\mathrm{OptVF}_d^k)$ are monotonically increasing in k. Furthermore, $w(\mathrm{OptE}^k) \leq w(\mathrm{OptEF}_d^k)$ for every d and $w(\mathrm{OptV}^k) \leq w(\mathrm{OptVF}_d^k)$ for every d.

We denote by MST a minimum-weight spanning tree of G.

1.2 Previous and Related Results

Without the triangle inequality, the problem of computing minimum-weight k-vertex-connected spanning subgraphs can be approximated with a factor of $O(\log k)$ [3], and the problem of computing minimum-weight k-edge-connected spanning subgraphs can be approximated with a factor of 2 [16]. However, no approximation at all seems to be possible without the triangle inequality if we ask for specific degrees. This follows from the inapproximability of non-metric TSP [25, Sect. 2.4].

With the triangle inequality, we obtain the same factor of 2 for k-edge-connected subgraphs of minimum weight without degree requirements [16]. For k-vertex-connected spanning subgraphs of minimum weight without degree constraints, Kortsarz and Nutov [17] gave a $\left(2 + \frac{k-1}{n}\right)$-approximation algorithm.

Cornelissen et al. [5] gave a 2.5 approximation for Min-dReg-2Edge for even d and a 3-approximation for Min-dReg-1Edge and Min-dReg-2Edge for all odd d.

Table 1. Overview of approximation ratios for Min-dReg-kEdge. Cases of odd k and even d are omitted as discussed in Sect. 1.1.

k	d	ratio	reference
$= 2$	$= 2$	1.5	same problem as TSP [25, Sect. 2.4]
$= 1$	odd	3	Cornelissen et al. [5]
≥ 3	$= k$	$2 + \frac{1}{k}$	Chan et al. [1]
≥ 2	$\geq k$, even	2.5	Theorem 2.13 ($k = 2$ by Cornelissen et al. [5])
≥ 2	$\geq k + 1$, odd	$4 - \frac{3}{k}$	Theorem 2.13

Table 2. Overview of approximation ratios for Min-dReg-kVertex.

k	d	ratio	reference
$\in \{1, 2\}$	$= 2$	1.5	same as problem as TSP [25, Sect. 2.4]
$\in \{1, 2, 3\}$	$= 3$		same as Min-dReg-kEdge
≥ 2	$= k$	$2 + \frac{k-1}{n} + \frac{1}{k}$	Chan et al. [1]
≥ 2	$= 2k - 1$	$5 + \frac{2k-2}{n} + \frac{2}{k}$	Theorem 3.2
≥ 2	$\geq 2k$	$5 + \frac{2k^n - 2}{n}$	Corollary 3.3

Also Min-kReg-kVertex and Min-kReg-kEdge admit constant factor approximations for all $k \geq 1$ [1]. We refer to Tables 1 and 2 for an overview.

Fukunaga and Nagamochi [8] considered the problem of finding a minimum-weight k-edge-connected spanning subgraph with given degree requirements. Different from the problem that we consider, they allow multiple edges between vertices. This considerably simplifies the problem as one does not have to take care to avoid multiple edges when constructing the approximate solution. For this relaxed variant of the problem, they obtain approximation ratios of 2.5 for even k and $2.5 + \frac{1.5}{k}$ for odd k if the minimum degree requirement is at least 2. We remark that, although an optimal solution with multiple edges cannot be heavier than an optimal solution without multiple edges, an approximation algorithm for the variant with multiple edges does not imply an approximation algorithm for the variant without multiple edges and vice versa.

In many cases of algorithms for network design with degree constraints, only bounds on the degrees are given or some violation of the degree requirements is allowed to simplify the problem. Fekete et al. [7] devised an approximation algorithm for the bounded-degree spanning tree problem. Given lower and upper bounds for the degree of every vertex, spanning trees can be computed that violate every degree constraint by at most 1 and whose weight is no more than the weight of an optimal solution [22]. Often, network design problems are considered as bicriteria problems, where the goal is to simultaneously minimize the total costs and the violation of the degree requirements [9, 10, 18–20]. In contrast, our goal is to meet the degree requirements exactly.

1.3   Our Contribution

We devise polynomial-time approximation algorithms for Min-dReg-kEdge (Sect. 2) and for Min-dReg-kVertex (Sect. 3). This answers an open question raised by Cornelissen et al. [5]. Our algorithms can be generalized to arbitrary degree sequences, as long as the minimum degree requirement is at least $2\lceil k/2 \rceil$ for edge connectivity or at least $2k - 1$ for vertex connectivity (Sect. 4).

We obtain an approximation ratio of $4 - \frac{3}{k}$ for Min-dReg-kEdge for odd $d \geq k + 1$, a ratio of 2.5 for Min-dReg-kEdge for even $d \geq k$, and an approximation ratio of about 5 for Min-dReg-kVertex for $d \geq 2k - 1$. The precise approximation ratios are summarized in Tables 1 and 2.

As far as we are aware, there do not exist any approximation results for the problem of finding subgraphs with exact degree requirements besides simple connectivity and 2-edge-connectivity [5]. The only exception that we are aware of is the work by Fukunaga and Nagamochi [8]. However, they allow multiple edges in their solutions, which seems to make the problem simpler to approximate.

The high-level ideas of our algorithms are as follows. For edge-connectivity, our initial idea was to iteratively increase the connectivity from $k - 1$ to k by considering the k-edge-connected components of the current solution and adding edges carefully. However, this does not work as k-edge-connected components are not guaranteed to exist in $(k - 1)$-edge-connected graphs. Instead, we introduce k-special components (Definition 2.1). By connecting the k-special components carefully, we can increase the edge-connectivity of the graph (Lemma 2.8). Every increase of the edge-connectivity costs at most $O(1/k)$ times the weight of the optimal solution (Lemma 2.10), yielding constant factor approximations for all k. Our algorithm for Min-dReg-kEdge generalizes the algorithm of Cornelissen et al. [5] to arbitrary k. A more careful analysis yields that already their algorithm achieves an approximation ratio of 2.5 for Min-dReg-2Edge also for odd d.

For vertex-connectivity, the idea is to compute a k-vertex-connected k-regular graph and a (possibly not connected) d-factor. We iteratively add edges from the k-vertex-connected graph to the d-factor while maintaining the degrees until we obtain a k-vertex-connected d-factor. This works for $d \geq 2k - 1$ (Lemma 3.1).

2   Edge-Connectivity

In this section, we present an approximation algorithm for Min-dReg-kEdge for all combinations of d and k, provided that $d \geq 2\lceil k/2 \rceil$. This means that the algorithm works for all $d \geq k$ with the only exception being the case of odd $d = k$.

The main idea of our algorithm is as follows: We start by computing a d-factor. Then we iteratively increase the connectivity as follows: First, we identify edges that we can safely remove without decreasing the connectivity. Second, we find edges that we can add in order to increase the connectivity while repairing the d-regularity that we have destroyed in the first step.

One might be tempted to use the k-edge-connected components of the d-factor in order to increase the edge-connectivity from $k - 1$ to k. The catch is that there need not be enough k-edge-connected components, and it is in fact possible to find $(k-1)$-edge-connected graphs that are d-regular with $d \geq k$ without any k-edge-connected component. To circumvent this problem, we introduce the notion of k-special components, which have the desired properties.

2.1 Graph-Theoretic Preparation

Different from the rest of the paper, the graph $G = (V, E)$ is not necessarily complete in this section. The following definition of k-special components is crucial for the whole Sect. 2.

Definition 2.1. *Let $k \in \mathbb{N}$, and let $G = (V, E)$ be a graph. We call $L \subseteq V$ a k-special component in G if $\mathrm{cut}_G(L) \leq k - 1$ and L is locally k-edge connected in G.*

For $k = 1$, the k-special components are the connected components of G. The 2-edge-connected components of a graph yield a tree with a vertex for every 2-edge-connected component and an edge between any 2-edge-connected components that are connected by an edge. The 2-special components of G correspond to the leaves of this tree.

Let us collect some facts about k-special components.

Lemma 2.2. *Let G have a minimum degree of at least k, and let L be a k-special component in G. Then $|L| \geq k + 1$.*

Lemma 2.3. *Let G be a graph. If L is a k-special component, then L is a maximal locally k-edge-connected component. If L and L' are k-special, then either $L = L'$ or $L \cap L' = \emptyset$.*

The following crucial lemma shows the existence of k-special components.

Lemma 2.4. *Let $k \geq 1$. Let $G = (V, E)$ be a $(k-1)$-edge-connected graph. Then every non-empty vertex set $X \subsetneq V$ either contains a k-special component or satisfies $\mathrm{cut}_G(X) \geq k$.*

The purpose of the next few lemmas is to show that we can always remove an edge from a k-special component without decreasing the connectedness of the whole graph. In the following, let $m = \lceil k/2 \rceil + 1$. It turns out that the graph induced by a k-special component contains a locally m-edge-connected component.

Lemma 2.5. *Let $k \geq 1$. Let $G = (V, E)$ be a $(k-1)$-edge-connected graph of minimum degree at least $2\lceil k/2 \rceil$, and let L be a k-special component of G. Then there exists an $X \subseteq L$ such that X is a locally m-edge-connected component in L and $|X| \geq k + 1$.*

The edges $\{u_i, v_i\}$ mentioned in the next lemma are the edges that we can safely remove. The resulting graph will remain $(k-1)$-edge-connected according to Lemma 2.7. The vertices u_i and v_i in the next lemma will be chosen from $X_i \subseteq L_i$, where X_i is a locally m-edge-connected component in L_i as in Lemma 2.5.

Lemma 2.6. *Let $k \geq 1$. Let $G = (V, E)$ be a $(k-1)$-edge-connected graph of minimum degree at least $2\lceil k/2 \rceil$. Let $L_1, \ldots, L_s$ be the k-special components of G. Then there exist vertices $u_i, v_i \in L_i$ for all $i \in \{1, \ldots, s\}$ such that the following properties are met:*

- *$\{u_i, v_i\} \in E$ for all i.*
- *$\{u_i, v_j\} \notin E$ for all $i \neq j$.*
- *There exist at least m edge-disjoint paths from u_i to v_i in the graph induced by L_i for every i.*

Lemma 2.7. *Let $G = (V, E)$ be a $(k-1)$-edge-connected graph of minimum degree at least $2\lceil k/2 \rceil$ with k-special components $L_1, \ldots, L_s$, and let $u_1, \ldots, u_s$ and $v_1, \ldots, v_s$ be chosen as in Lemma 2.6. Let $Q = \{\{u_i, v_i\} \mid 1 \leq i \leq s\}$. Then $G - Q$ is $(k-1)$-edge-connected.*

By removing the edges $\{u_i, v_i\} \in Q$ and adding the edges $\{u_i, v_{i+1}\} \in S$, we construct a k-edge-connected graph from the $(k-1)$-edge-connected graph G according to the following lemma.

Lemma 2.8. *Let $G = (V, E)$ be a $(k-1)$-edge-connected graph of minimum degree at least $2\lceil k/2 \rceil$ with k-special components $L_1, \ldots, L_s$, and let $u_1, \ldots, u_s$ and $v_1, \ldots, v_s$ be chosen as in Lemma 2.6. Let $Q = \{\{u_i, v_i\} \mid 1 \leq i \leq s\}$, and let $S = \{\{u_i, v_{i+1}\} \mid 1 \leq i \leq s\}$, where arithmetic is modulo s.*
Then the graph $\tilde{G} = G - Q + S$ is k-edge-connected.

To conclude this section, we remark that the k-special components of a graph can be found in polynomial-time: local k-edge-connectedness can be tested in polynomial time. Thus, we can find locally k-edge-connected components in polynomial time. Since k-special components are maximal locally k-edge-connected components, we just have to compute a partition of the graph into locally k-edge-connected components and check whether less than k edges leave such a component. Therefore, the sets L_i and $X_i \subseteq L_i$ as well as the vertices u_i and v_i with the properties as in Lemmas 2.5 and 2.6 can be computed in polynomial time.

2.2   Algorithm and Analysis

Our approximation algorithm for Min-dReg-kEdge (Algorithm 1) starts with an ℓ-edge-connected d-factor F_ℓ. How we choose ℓ and compute F_ℓ depends on the parity of k, but it is possible that improved approximation algorithms for certain small k lead to other initializations. (If k is even, we use $\ell = 0$ and $F_0 = \mathrm{OptF}_d$. If k is odd, we use $\ell = 2$ and approximate a 2-edge-connected d-factor F_2 using the algorithm of Cornelissen et al. [5].)

Then it iteratively uses a subroutine (Algorithm 2) that increases the connectivity. To increase the connectivity, we compute a TSP tour (line 3). We do this using Christofides' algorithm [25, Sect. 2.4], which achieves an approximation ratio of 1.5.

---

input : undirected complete graph $G = (V, E)$, edge weights w, integers $k \geq 3$,
         $d \geq 2\lceil k/2 \rceil$

output: k-edge-connected d-factor R of G

1 compute a minimum-weight ℓ-edge-connected d-factor F_ℓ (or an approximation)
2 **for** $p \leftarrow \ell + 1, \ldots, k$ **do**
3   **if** F_{p-1} *is not p-edge-connected* **then**
4 | apply Algorithm 2 to obtain F_p
5   **else**
6 | $F_p \leftarrow F_{p-1}$
7   **end**
8 **end**
9 $R \leftarrow F_k$

Algorithm 1. Approximation algorithm for Min-dReg-kEdge.

---

input : undirected complete graph $G = (V, E)$, edge weights w, integer $p \geq 1$,
         $(p - 1)$-edge-connected subgraph F_{p-1} of G with minimum degree at
         least $p + 1$

output: p-edge-connected subgraph F_p of G with the same degree at every
         vertex as F_{p-1}

1 find the p-special components of F_{p-1}; let $L_1, \ldots, L_s$ be these p-special
   components
2 find vertices $u_i, v_i \in L_i$ for all $i \in \{1, \ldots, s\}$ with the properties stated in
   Lemma 2.6; $Q \leftarrow \{\{u_i, v_i\} \mid 1 \leq i \leq s\}$
3 compute a TSP tour T on V using Christofides' algorithm
4 take shortcuts to obtain a tour T' on $u_1, \ldots, u_s$ (without loss of generality in
   this order)
5 $S \leftarrow \{\{u_i, v_{i+1}\} \mid 1 \leq i \leq s\}$ (arithmetic modulo s)
6 $F_p \leftarrow F_{p-1} - Q + S$

Algorithm 2. Increasing the edge-connectivity of a graph by 1 while maintaining d-regularity.

We analyze correctness and approximation ratio using a series of lemmas.

Lemma 2.9. *Let $k \geq 1$ be arbitrary, and let $p \in \{\ell, \ell + 1, \ldots, k\}$. Let F_p be computed by Algorithm 1. Then F_p is d-regular and p-edge-connected.*

In order to analyze the approximation ratio and to achieve a constant approximation for all k, we exploit a result that Fukunaga and Nagamochi [8] attributed to Goemans and Bertsimas [12] and Wolsey [26].

Lemma 2.10 (Fukunaga, Nagamochi [8, Theorem 2]). *Let T be the TSP tour obtained from Christofides' algorithm. Then $w(T) \leq \frac{3}{k} \cdot w(\mathrm{OptE}^k)$.*

A consequence of Lemma 2.10 are the following two statements, which we need to analyze the approximation ratio.

Lemma 2.11. *If, in Algorithm 1, we enter line 4 and call Algorithm 2, then*

$$w(F_p) \leq \frac{3}{k} \cdot w(\mathrm{OptEF}_d^k) + w(F_{p-1}).$$

Lemma 2.12. *If Algorithm 1 calls Algorithm 2 q times, then*

$$w(F_k) \leq \frac{3q}{k} \cdot w(\mathrm{OptEF}_d^k) + w(F_\ell).$$

Theorem 2.13. *For $k \geq 2$ and $d \geq 2\lceil k/2 \rceil$, Algorithm 1 is a polynomial-time approximation algorithm for* Min-dReg-kEdge. *It achieves an approximation ratio of 2.5 for even d and an approximation ratio of $4 - \frac{3}{k}$ for odd d.*

Algorithm 1 works also for the case of even $d = k$, but there exists already an approximation algorithm with a ratio of $2 + \frac{1}{k}$ for this special case [1]. Note that the proof of Theorem 2.13 does not cover the case of odd d and $k = 1$, but it is already known that this case can be approximated with a factor of 3 [5].

3   Vertex Connectivity

In this section, we consider Min-dReg-kVertex for $d \geq 2k - 1$. The basis of the algorithm (Algorithm 3) is the following: Assume that we have a k-vertex connected k-factor H and a d-factor F that lacks k-vertex-connectedness. Then we iteratively add edges from H to F to make F k-vertex-connected as well. More precisely, we try to add an edge $e \in H \setminus F$ to increase the connectivity of F. To maintain that F is d-regular, we have to add another edge and remove two edges of F. If, in the course of this process, we never have to remove an edge of H from F, then the algorithm terminates with a k-vertex-connected d-regular graph.

In Algorithm 3, the initial d-factor OptF_d can be computed in polynomial time (line 1). Kortsarz and Nutov showed that we can compute a k-vertex-connected spanning subgraph K whose total weight is at most a factor of $2 + \frac{k-1}{n}$ larger than the weight of a k-vertex-connected graph of minimum weight (line 2). Chan et al. [1] devised an algorithm that turns k-vertex-connected graphs K into k-regular k-vertex-connected graphs H at the expense of an additive $w(\mathrm{OptV}^k)/k$.

With this initialization, we iteratively add edges from H to F while maintaining d-regularity of F. This works as long as d is sufficiently large according to the following lemma. We parametrize the maximum degree by ℓ in order to be able to get a slight improvement for larger d (Corollary 3.3).

Lemma 3.1. *Let $k, \ell \geq 2$ and $d \geq k + \ell - 1$. Let $G = (V, E)$ be an undirected complete graph. Let F be a d-factor of G, and let H be a k-vertex-connected graph subgraph of G that has a maximum degree of at most ℓ. Assume that F is not k-vertex-connected.*

Then there exists an edge $e = \{u_1, u_2\} \in H \setminus F$ such that u_1 and u_2 are not connected via k vertex-disjoint paths in F. Furthermore, given such an edge $e = \{u_1, u_2\}$, there exist vertices $v_1, v_2 \in V$ with $v_1 \neq v_2$ and the following properties:

---

input : undirected complete graph $G = (V, E)$, edge weights w, integers $k \geq 2$,
    $d \geq 2k - 1$
output: k-vertex-connected d-factor R of G
1 $F \leftarrow \text{OptF}_d$
2 approximate a k-vertex connected graph K using the algorithm of Kortsarz and
   Nutov [17]
3 compute a k-vertex-connected k-factor H from K using the algorithm of Chan
   et al. [1]
4 **while** F is not k-vertex-connected **do**
5   select an edge $e = \{u_1, u_2\} \in H \setminus F$ such that u_1 and u_2 are not connected
     by k vertex-disjoint paths in F
6   choose vertices v_1, v_2 with $\{u_1, v_1\}, \{u_2, v_2\} \in F \setminus H$ and $\{v_1, v_2\} \notin F$
7      $F \leftarrow \big(F \setminus \{\{u_1, v_1\}, \{u_2, v_2\}\}\big) \cup \{\{u_1, u_2\}, \{v_1, v_2\}\}$
8 **end**
9 $R \leftarrow F$

Algorithm 3. Approximation algorithm for Min-dReg-kVertex for $d \geq 2k - 1$.

1. $\{u_1, v_1\}, \{u_2, v_2\} \in F \setminus H$.
2. $\{v_1, v_2\} \notin F$.

With this lemma, we can prove the main result of this section.

Theorem 3.2. *For $k, d \in \mathbb{N}$ with $k \geq 2$ and $d \geq 2k - 1$, Algorithm 3 is a polynomial-time approximation algorithm for Min-dReg-kVertex with an approximation ratio of $5 + \frac{2k-2}{n} + \frac{2}{k}$.*

Algorithm 3 also works for $k = 1$, but there already exist better approximation algorithms for this case (see Table 2). With the slightly stronger assumption $d \geq 2k$, we can get a slightly better approximation ratio.

Corollary 3.3. *For $k, d \in \mathbb{N}$ with $k \geq 2$ and $d \geq 2k$, there exists a polynomial-time approximation algorithm for Min-dReg-kVertex with an approximation ratio of $5 + \frac{2k-2}{n}$.*

4   Generalization to Arbitrary Degree Sequences

Both algorithms of Sects. 2 and 3 do not exploit d-regularity, but only that the degree of each vertex is at least d. Thus, we immediately get approximation algorithms for Min-dGen-kEdge and Min-dGen-kVertex, where we get a degree requirement of at least d for each vertex.

For k-edge-connectedness, we require that the minimum degree requirement is at least $2\lceil k/2 \rceil$.

Theorem 4.1. *For $k \geq 2$, Min-$(2\lceil \frac{k}{2} \rceil)$Gen-$k$Edge can be approximated in polynomial time with an approximation ratio of $4 - \frac{3}{k}$.*

For k-vertex-connectivity, we require that the minimum degree requirement is at least $2k - 1$. (For minimum degree at least $2k$, we get a small improvement similarly to Corollary 3.3.)

Theorem 4.2. *For $k \geq 2$, Min-$(2k - 1)$Gen-kVertex can be approximated in polynomial time with an approximation ratio of $5 + \frac{2k-2}{n} + \frac{2}{k}$.*
  Min-$(2k)$Gen-kVertex can be approximated in polynomial time with an approximation ratio of $5 + \frac{2k-2}{n}$.

5   Conclusions and Open Problems

We conclude this paper with two questions for further research.

First, for edge-connectivity, we require $d \geq 2\lceil k/2 \rceil$. Since there exists an approximation algorithm for Min-kReg-kEdge (for $k \geq 2$) [1], the only case for which it is unknown if a constant factor approximation algorithm exists is the generalized problem Min-kGen-kEdge for odd values of k. We are particularly curious about approximation algorithms for Min-1Gen-1Edge, where we want to find a cheap connected graph with given vertex degrees. To get such algorithms, vertices with degree requirement 1 seem to be bothersome. (This seems to be a more general phenomenon in network design, as, for instance, the approximation algorithms by Fekete et al. [7] for bounded-degree spanning trees and by Fukunaga and Nagamochi [8] for k-edge-connected subgraphs with multiple edges both require that the minimum degree requirement is at least 2.) Still, we conjecture that constant factor approximation algorithms exist for these problems as well.

Second, we would like to see constant factor approximation algorithms for Min-dReg-kVertex for the case $k + 1 \leq d \leq 2k - 2$ and for the general problem Min-dGen-kVertex for $k \leq d \leq 2k - 2$. We conjecture that constant factor approximation algorithms exist for these problems.

References

1. Chan, Y.H., Fung, W.S., Lau, L.C., Yung, C.K.: Degree bounded network design with metric costs. SIAM J. Comput. **40**(4), 953–980 (2011)
2. Cheah, F., Corneil, D.G.: The complexity of regular subgraph recognition. Discrete Appl. Math. **27**(1–2), 59–68 (1990)
3. Cheriyan, J., Vempala, S., Vetta, A.: An approximation algorithm for the minimum-cost k-vertex connected subgraph. SIAM J. Comput. **32**(4), 1050–1055 (2003)
4. Cheriyan, J., Vetta, A.: Approximation algorithms for network design with metric costs. SIAM J. Discrete Math. **21**(3), 612–636 (2007)
5. Cornelissen, K., Hoeksma, R., Manthey, B., Narayanaswamy, N.S., Rahul, C.S.: Approximability of connected factors. In: Kaklamanis, C., Pruhs, K. (eds.) WAOA 2013. LNCS, vol. 8447, pp. 120–131. Springer, Heidelberg (2014)
6. Czumaj, A., Lingas, A.: Minimum k-connected geometric networks. In: Kao, M.Y. (ed.) Encyclopedia of Algorithms, pp. 536–539. Springer, Heidelberg (2008)

7. Fekete, S.P., Khuller, S., Klemmstein, M., Raghavachari, B., Young, N.E.: A network-flow technique for finding low-weight bounded-degree spanning trees. J. Algorithms **24**(2), 310–324 (1997)
8. Fukunaga, T., Nagamochi, H.: Network design with edge-connectivity and degree constraints. Theor. Comput. Syst. **45**(3), 512–532 (2009)
9. Fukunaga, T., Nagamochi, H.: Network design with weighted degree constraints. Discrete Optim. **7**(4), 246–255 (2010)
10. Fukunaga, T., Ravi, R.: Iterative rounding approximation algorithms for degree-bounded node-connectivity network design. In: Proceedings of the 53rd Annual Symposium on Foundations of Computer Science (FOCS), pp. 263–272. IEEE Computer Society (2012)
11. Garey, M.R., Johnson, D.S.: Computers and Intractability: A Guide to the Theory of NP-Completeness. W. H. Freeman and Company, New York (1979)
12. Goemans, M.X., Bertsimas, D.: Survivable networks, linear programming relaxations and the parsimonious property. Math. Program. **60**, 145–166 (1993)
13. Kammer, F., Täubig, H.: Connectivity. In: Brandes, U., Erlebach, T. (eds.) Network Analysis. LNCS, vol. 3418, pp. 143–177. Springer, Heidelberg (2005)
14. Khandekar, R., Kortsarz, G., Nutov, Z.: On some network design problems with degree constraints. J. Comput. Syst. Sci. **79**(5), 725–736 (2013)
15. Khuller, S., Raghavachari, B.: Graph connectivity. In: Kao, M.Y. (ed.) Encyclopedia of Algorithms, pp. 371–373. Springer, Heidelberg (2008)
16. Khuller, S., Vishkin, U.: Biconnectivity approximations and graph carvings. J. ACM **41**(2), 214–235 (1994)
17. Kortsarz, G., Nutov, Z.: Approximating node connectivity problems via set covers. Algorithmica **37**(2), 75–92 (2003)
18. Lau, L.C., Naor, J., Salavatipour, M.R., Singh, M.: Survivable network design with degree or order constraints. SIAM J. Comput. **39**(3), 1062–1087 (2009)
19. Lau, L.C., Singh, M.: Additive approximation for bounded degree survivable network design. SIAM J. Comput. **42**(6), 2217–2242 (2013)
20. Lau, L.C., Zhou, H.: A unified algorithm for degree bounded survivable network design. In: Lee, J., Vygen, J. (eds.) IPCO 2014. LNCS, vol. 8494, pp. 369–380. Springer, Heidelberg (2014)
21. Lovász, L., Plummer, M.D.: Matching Theory, North-Holland Mathematics Studies, vol. 121. Elsevier (1986)
22. Singh, M., Lau, L.C.: Approximating minimum bounded degree spanning trees to within one of optimal. J. ACM **62**(1), 1:1–1:19 (2015)
23. Tutte, W.T.: A short proof of the factor theorem for finite graphs. Can. J. Math. **6**, 347–352 (1954)
24. West, D.B.: Introduction to Graph Theory, 2nd edn. Prentice-Hall, Upper Saddle River (2001)
25. Williamson, D.P., Shmoys, D.B.: The Design of Approximation Algorithms. Cambridge University Press, New York (2011)
26. Wolsey, L.A.: Heuristic analysis, linear programming and branch and bound. In: Rayward-Smith, V.J. (ed.) Combinatorial Optimization II, Mathematical Programming Studies, vol. 13, pp. 121–134. Springer, Heidelberg (1980)

Improved Approximation Algorithms for Unsplittable Flow on a Path with Time Windows

Fabrizio Grandoni, Salvatore Ingala, and Sumedha Uniyal [✉]

IDSIA, University of Lugano, Lugano, Switzerland
{fabrizio,salvatore,sumedha}@idsia.ch

Abstract. In the well-studied *Unsplittable Flow on a Path* problem (UFP), we are given a path graph with edge capacities. Furthermore, we are given a collection of n tasks, each one characterized by a sub-path, a weight, and a demand. Our goal is to select a maximum weight subset of tasks so that the total demand of selected tasks using each edge is upper bounded by the corresponding capacity. Chakaravarthy et al. [ESA'14] studied a generalization of UFP, bagUFP, where tasks are partitioned into bags, and we can select at most one task per bag. Intuitively, bags model jobs that can be executed at different times (with different duration, weight, and demand). They gave a $O(\log n)$ approximation for bagUFP. This is also the best known ratio in the case of uniform weights. In this paper we achieve the following main results:

- We present an LP-based $O(\log n/ \log \log n)$ approximation for bagUFP. We remark that, prior to our work, the best known integrality gap (for a non-extended formulation) was $O(\log n)$ even in the special case of UFP [Chekuri et al., APPROX'09].
- We present an LP-based $O(1)$ approximation for uniform-weight bagUFP. This also generalizes the integrality gap bound for uniform-weight UFP by Anagnostopoulos et al. [IPCO'13].
- We consider a relevant special case of bagUFP, twUFP, where tasks in a bag model the possible ways in which we can schedule a job with a given processing time within a given time window. We present a QPTAS for twUFP with quasi-polynomial demands and under the Bounded Time-Window Assumption, i.e. assuming that the time window size of each job is within a constant factor from its processing time. This generalizes the QPTAS for UFP by Bansal et al. [STOC'06].

1 Introduction

In the well-studied *Unsplittable Flow on a Path* problem (UFP) we are given a path graph $G = (V, E)$, $V = \{0, 1, \ldots, m\}$, with positive integer edge capacities $\{u_e\}_{e \in E}$ and a collection T of n tasks. Each task $i \in T$ is associated with a weight $w_i \in \mathbb{N}^+$, a demand $d_i \in \mathbb{N}^+$, and a subpath P_i between nodes s_i and t_i.

This work is partially supported by the ERC StG project NEWNET no. 279352.

L. Sanità and M. Skutella (Eds.): WAOA 2015, LNCS 9499, pp. 13–24, 2015.
DOI: 10.1007/978-3-319-28684-6_2

Let $T_e = \{i \in T : e \in P_i\}$ be the tasks *containing* edge e. Our goal is to select a subset of tasks $T' \subseteq T$ of maximum total weight $w(T') := \sum_{i \in T'} w_i$ so that, for each edge e, the total demand $d_e(T') := \sum_{i \in T' \cap T_e} d_i$ of selected tasks using that edge is upper bounded by the corresponding capacity u_e. Intuitively, edge capacities model a given resource whose amount varies over a given time interval (in a discrete fashion), and tasks demand for some amount of that resource. In particular, the length of each subpath can be interpreted as a processing time. By standard reductions [4,8], we can assume that $m \leq 2n$ and all edge capacities are distinct.

UFP is strongly NP-hard [14]. Anagnostopoulos et al. [2] recently gave the current best $2 + \varepsilon$ approximation for the problem[1], improving on [5,8]. This matched a previously known [13] approximation for UFP under the No-Bottleneck Assumption (NBA), i.e. assuming that the largest demand is upper bounded by the smallest capacity. This matched also the best known approximation for the uniform-capacity case [9].

The *UFP with Bags* problem (bagUFP) is the generalization of UFP where tasks are partitioned into a set of h bags $J = \{\mathcal{B}_1, \ldots, \mathcal{B}_h\}$, and we have the extra constraint that at most one task per bag can be selected. Intuitively, bags model *jobs* that we can execute at different points of time (and at each such time one has a different demand, weight, and processing time). This problem is APX-hard even in the case of unit demands and capacities [16]. Chakaravarthy et al. [10] recently gave the current best $O(\log n)$ approximation for bagUFP. The approximation factor remains the same in the case of uniform weights. The same authors also presented a $O(1)$ approximation under NBA.

1.1 Our Contribution.

In this paper we present an improved approximation for bagUFP (see Section 3). In the special case of uniform weights, we can reduce the approximation factor down to a constant (see Section 4).

Theorem 1. *There is an expected $O(\log n / \log \log n)$ approximation for bagUFP.*

Theorem 2. *There is an $O(1)$ approximation for uniform-weight bagUFP.*

Both our results are LP-based, and exploit a refined LP for bagUFP which is inspired by the work on UFP in [1,8]. In more detail, let us define a task *large* if it uses more than one half of the capacity of some edge along its subpath, and let T_{large} be the large tasks. Bonsma et al. [8] introduced a geometric interpretation of *large* tasks. They associate to each $i \in T_{large}$ an axis-parallel rectangle R_i in the 2D plane with top-left corner (s_i, b_i) and bottom-right corner $(t_i, b_i - d_i)$, where $b_i = \min_{e \in P_i}\{u_e\}$ is the bottleneck capacity of task i. We call this set of rectangles $\mathcal{R}$ the *top-drawn* representation of T_{large}. In [8, Lemma 13] it is shown that in any feasible UFP (hence bagUFP) solution at most 4 corresponding rectangles can overlap at a given point (intuitively, those large tasks induce

---

[1] Unless differently stated, ε denotes an arbitrarily small positive constant parameter. Where needed, we also assume that $1/\varepsilon$ is integral and sufficiently large.

an *almost independent* set of rectangles). This insight was later used by Anagnostopoulos et al. [1]. Consider the grid induced by the horizontal and vertical lines containing the rectangle sides. Let $\mathcal{P}$ be the set of (*representative*) $O(n^2)$ middle points of the (positive area) cells of this grid. Note that any subset of rectangles that share a positive size area, will overlap on some point in $\mathcal{P}$. The authors consider the following (non-extended[2]) LP relaxation for UFP:

$$\max \quad \sum_{i \in T} w_i x_i \qquad\qquad\qquad\qquad (LP_{UFP+})$$

$$\text{s.t.} \quad \sum_{i:e \in P_i} d_i x_i \leq u_e \qquad\qquad \forall e \in E \qquad\qquad (1)$$

$$\sum_{i \in T_{large}:p \in R_i} x_i \leq 4 \qquad\qquad \forall p \in \mathcal{P} \qquad\qquad (2)$$

$$x_i \geq 0 \qquad\qquad\qquad\qquad \forall i \in T$$

We call the constraints of type (1) and (2) *capacity* and *rectangle constraints*, respectively. The authors show that this relaxation has $O(1)$ integrality gap in the case of uniform weights. Note that one can preprocess the instance so that the weights range between 1 and $O(n/\varepsilon)$ while losing a factor $1+\varepsilon$ in the approximation. By partitioning tasks in $O(\log(n/\varepsilon))$ classes of almost uniform weight, one obtains that LP_{UFP+} has $O(\log n)$ integrality gap for general weights[3].

In this paper we consider the LP relaxation $LP_{bagUFP+}$ for bagUFP which is obtained from LP_{UFP+} by adding the following *bag constraints*:

$$\sum_{i \in \mathcal{B}_j} x_i \leq 1 \qquad \forall \mathcal{B}_j \in J.$$

The standard LP relaxation LP_{bagUFP} for bagUFP is obtained from bagUFP$^+$ by removing the rectangle constraints. We show that $LP_{bagUFP+}$ has constant integrality gap in the uniform weight case, and integrality gap $O(\log n / \log \log n)$ in the general case. In particular, for the uniform-weight case we can adapt the analysis in [1], while for the general case we can generalize the rounding procedure of Chan and Har-Peled [11] for the maximum independent set of rectangles problem. Note that the latter result slightly improves the best integrality gap even in the case of UFP (for a compact, non-extended LP relaxation).

We also study a relevant special case of bagUFP (and generalization of UFP), that we name *UFP with Time Windows* (twUFP). Here we are given a capacitated path graph and a collection of *jobs*, where each job j is characterized by a weight, a demand, a (positive, integer) *processing time* τ_j and a *time window* W_j (i.e. a subpath between given nodes s_j and t_j). For each possible node σ_i (*starting time*) so that $s_j \leq \sigma_i \leq t_j - \tau_j$, we define a task i with the same weight and demand as j, and whose subpath P_i has endpoint $s_i = \sigma_i$ and $t_i = \sigma_i + \tau_i$. The tasks corresponding to the same job j define a bag $\mathcal{B}_j$. Intuitively, tasks in $\mathcal{B}_j$ describe the possible ways in which we can process job j within its time window. Our goal is to compute a maximum weight solution for the resulting bagUFP instance. We believe that in practice several instances of bagUFP

---

[2] By non-extended we mean that it contains only decision variables for tasks. In the same paper the authors present an extended formulation with $O(1)$ integrality gap.

[3] The same gap is proved by Chekuri et al. [12]. The authors claim a $O(\log^2 n)$ gap, and then refine it to $O(\log n)$ in an unpublished manuscript.

are indeed instances of twUFP. The best-known approximation for twUFP is also $O(\log n)$, where n is the number of tasks ($O(\log n / \log \log n)$ considering Theorem 1). In particular, the approach in [10] does not seem to benefit from the special structure of twUFP, nor does the approach from Theorem 1.

We present a QPTAS[4] for twUFP under the following *Bounded Time-Window Assumption* (BTWA): for any job j, $(t_j - s_j)/\tau_j \leq C = O(1)$ (in words, the ratio between the time window size and the processing time is bounded by some constant). Our result generalizes the QPTAS for UFP by Bansal et al. [4]. Here, similarly to [4], we assume[5] that demands are quasi-polynomially bounded in n.

Theorem 3. *There is a QPTAS for twUFP under BTWA and assuming that demands are quasi-polynomially bounded in n.*

Indeed, our QPTAS generalizes to the special case of bagUFP where tasks in the same bag have the same demand and weight (under the natural generalization of BTWA, that is, under the assumption that the processing time of any task i contained in bag $\mathcal{B}_j$ is at most a constant factor shorter than the length of the bag $\max_{i \in \mathcal{B}_j} t_i - \min_{i \in \mathcal{B}_j} s_i$). Note that bagUFP is APX-hard, therefore there is not much hope for a PTAS for it. In contrast, our result provides an evidence that twUFP might be an easier problem, at least under BTWA. It is unclear to us whether the general case of twUFP is as hard to approximate as bagUFP.

In order to understand our contribution, it is convenient to sketch how the QPTAS for UFP in [4] works. Let us consider the tasks OPT_{mid} in the optimum solution OPT that use the middle edge e_{mid}. The authors show how to define a *capacity profile* u_{mid}, *dominated* by the demand of OPT_{mid}, which has a quasi-polynomial number of *steps*, and such that there is a feasible solution APX_{mid} for capacities u_{mid} of cost close to OPT_{mid} and which can be computed in QPT. Thus one can *guess* u_{mid}, compute APX_{mid}, and branch on a left and right subproblem (where capacities are decreased by u_{mid}, and we consider only tasks fully contained to the left/right of e_{mid}).

This approach does not work for twUFP since a time window might be split by e_{mid}. In that case the left and right subproblems are not independent any more (in particular, we cannot select two tasks from the same bag, one from the left subproblem and the other from the right one). To circumvent this problem, we exploit the *randomized dissection* technique by Grandoni and Rothvoß for the related *Highway* problem [15]. We evenly split the path into a random constant number of intervals, and iterate the process on each such interval. With probability close to one, the time window of each job j is fully contained in some interval I of the dissection, and at the same time none of its tasks $\mathcal{B}_j$ is fully contained in a subinterval of I (here we need the BTWA). Thus we can define a

---

[4] We recall that a *Quasi-Polynomial-Time Approximation Scheme* (QPTAS) is an algorithm that, given a constant parameter $\varepsilon > 0$, computes a $1 + \varepsilon$ approximation in *Quasi-Polynomial Time* (QPT), i.e. in time $2^{poly \log(s)}$ where s is the input size.

[5] We remark that Batra et al. [7] recently managed to remove this assumption on the demands for UFP. Their approach does not seem to be compatible with our randomized dissection technique (at least not trivially).

capacity reservation that combines a constant number of capacity profiles, and use a proper algorithm (rather different from [4], due to bag constraints) to compute a good approximation for the considered jobs. We can then branch on the subproblems induced by the subintervals.

Omitted proofs are provided in the full version of this paper.

2   A QPTAS for the Bounded Time-Window Case

In this section we present a QPTAS for twUFP under BTWA, and assuming that the largest demand D_{max} is quasi-polynomially bounded in the number of tasks n. By standard tricks, while losing only a factor $1 + \varepsilon$ in the approximation factor, we can assume that weights range between 1 and $O(n/\varepsilon)$.

A *capacity reservation* r is simply a collection of edge capacities $\{r_e\}_{e \in E}$ with $r_e \leq u_e$ for all edges e. A solution is feasible w.r.t. r if it respects the capacity constraints induced by r. We say that r has k-steps if, scanning edges from left to right, the value of their capacity changes at most k times. For another capacity reservation r', we say that r *dominates* r' if $r_e \geq r'_e$ for each edge e. For a set of tasks S, we say that r *is dominated by the demand of S* if $r_e \leq \sum_{i \in S : e \in P_i} d_i$ for each edge e.

The following technical lemma is similar in spirit to results in [4].

Lemma 1. *Let S be a collection of at least $2/\varepsilon^3$ tasks using a given edge e, and with demand in $[D, (1 + \varepsilon)D)$ and weight in $[W, (1 + \varepsilon)W)$. Then there exists a capacity reservation r and a set of tasks $R \subseteq S$ such that: (1) r is dominated by the demand of S; (2) r has $O(1/\varepsilon^2)$ steps and its entries are integer multiples of $(1 + \varepsilon)D$; (3) R is feasible for r, even if the paths of its tasks are expanded to the left/right to reach the closest edge before a change of capacity in r and their demand is increased to $(1 + \varepsilon)D$; (4) $w(R) \geq (1 - O(\varepsilon))w(S)$.*

Next lemma will be used to partition the input problem into a quasi-polynomial number of subproblems. For a given set of edges F, let J_F be the set of jobs j such that each task in $\mathcal{B}_j$ contains some edge in F, and let $T_F = \cup_{j \in J_F} \mathcal{B}_j$. Note that containing an edge in F is not sufficient for a task i to be in T_F. We define $T_{\bar{F}} = T \setminus T_F$.

Lemma 2. *Consider a twUFP instance with optimal solution OPT, and let F be a subset of $O(1)$ edges. There exists a QPT algorithm that generates a set $\mathcal{U}_F$ of capacity reservations r_F, and a feasible solution $APX_F \subseteq T_F$ for each such r_F such that, for at least one such pair $\{r^*, APX^*\}$, $APX := (OPT \cap T_{\bar{F}}) \cup APX^*$ is a feasible twUFP solution and $w(APX) \geq (1 - O(\varepsilon))w(OPT)$.*

Proof. Let $OPT_F := OPT \cap T_F$ and $OPT_{\bar{F}} := OPT \cap T_{\bar{F}}$. We first show how to construct a capacity reservation r^* which is dominated by the demand of OPT_F. Let $T^{f,a,b}$ be the class of tasks $i \in T_F$ with $d_i \in [(1 + \varepsilon)^a, (1 + \varepsilon)^{a+1})$

and $w_i \in [(1+\varepsilon)^b, (1+\varepsilon)^{b+1})$ for $a, b \in \mathbb{N}$, and such that f is the leftmost edge in $P_i \cap F$. Observe that there are $O_\varepsilon(\log n \log D_{max})$ (non-empty) such classes. Define $OPT^{f,a,b} := OPT \cap T^{f,a,b}$. Suppose that $|OPT^{f,a,b}| \geq 2/\varepsilon^3$. Then we apply Lemma 1 with $S = OPT^{f,a,b}$ and $e = f$, hence obtaining a capacity reservation $r^{f,a,b}$ and a solution $R^{f,a,b}$. Otherwise, we simply let $R^{f,a,b} = OPT^{f,a,b}$ and $r^{f,a,b}$ be the total demand of $R^{f,a,b}$. Let $r^* = \sum_{f,a,b} r^{f,a,b}$. Observe that r^* has $O_\varepsilon(|F| \log n \log D_{max})$ steps, and each entry of r^* is obtained from the total demand of $O(|F|/\varepsilon^3)$ tasks plus an integer multiple of $(1+\varepsilon)^{a+1}$ for $O_\varepsilon(\log D_{max})$ possible values of a. Therefore in QPT we can enumerate a set $\mathcal{U}_F$ of capacity reservations that includes r^*.

Our algorithm constructs (in QPT) a feasible solution for each capacity reservation in $\mathcal{U}_F$. For the sake of simplicity, we next focus on the solution APX^* corresponding to the reservation r^* described before. Note that, since r^* is dominated by the demand of OPT_F, $APX^* \cup OPT_{\bar{F}}$ has to satisfy the capacity constraints. Furthermore, the bags of tasks in $OPT_{\bar{F}}$ are disjoint from the bags of tasks in T_F by definition (hence also bag constraints are satisfied). We will later show that $w(APX^*) \geq (1 - O(\varepsilon))w(OPT_F)$. The claim follows.

Let us focus on a given pair (a, b). We guess[6] the set $F^{a,b}_{few}$ of all the edges $f \in F$ such that $|OPT^{f,a,b}| < 2/\varepsilon^3$, and the corresponding tasks $APX^{f,a,b} := OPT^{f,a,b} = R^{f,a,b}$ with demand $r^{f,a,b}$. The corresponding jobs are removed from the instance. Let $F^{a,b}_{many} := F \setminus F^{a,b}_{few}$. For any $f \in F^{a,b}_{many}$, we guess the capacity reservation $r^{f,a,b}$.[7] Note that this reservation has $O(1/\varepsilon^2)$ steps, and its entries are integer multiples of $(1+\varepsilon)D$, $D = (1+\varepsilon)^a$. We expand all the tasks in $T^{f,a,b}$ to the left/right till the closest edge before a change in the capacity of $r^{f,a,b}$ and increase their demand to $(1+\varepsilon)D$.

Next we consider the bagUFP instance induced by rounded tasks $T^{a,b}_{many} := \cup_{f \in F^{a,b}_{many}} T^{f,a,b}$, with edge capacities given by $r^{a,b}_{many} := \sum_{f \in F^{a,b}_{many}} r^{f,a,b}$. Observe that all the tasks of a remaining job are considered in the same such instance[8]. We also remark that $R^{a,b}_{many} := \cup_{f \in F^{a,b}_{many}} R^{f,a,b}$ is a feasible solution to this bagUFP instance by construction. We also remark that $r^{a,b}_{many}$ has $O(|F|/\varepsilon^2)$ steps, hence by contracting edges one obtains an equivalent bagUFP instance with a constant number of edges.

We next show how to compute the optimal solution $APX^{a,b}_{many}$ for this bagUFP instance via dynamic programming. Let us sort the considered $h^{a,b}_{many}$ jobs arbitrarily. In our dynamic program we have a table entry (h', r') for each $h' = 1, \ldots, h^{a,b}_{many}$ and for each feasible capacity reservation r' dominated by $r^{a,b}_{many}$ and whose capacities are non-negative integer multiples of $(1+\varepsilon)D$. Note that there is a polynomial number of table entries. The value $DP(h', r')$ of this entry will be set to the maximum weight of a feasible bagUFP for r' using tasks from the first h' jobs only. Table entries are filled in for increasing values of h'.

---

[6] Throughout this paper, by guessing we mean trying all the possibilities.

[7] In the guessing we of course guarantee that $r^* = \sum_{f,a,b} r^{f,a,b}$.

[8] Here we exploit a property of twUFP not satisfied by bagUFP.

It is easy to compute the values $DP(1, r')$ (base case). For any $h' > 1$, one has[9]

$$DP(h', r') = \max\{DP(h'-1, r'), \max_{i \in \mathcal{B}_j}\{w_j + DP(h'-1, r_i')\}\},$$

where r_i' is obtained from r' by subtracting the demand of task i. The desired solution $APX_{many}^{a,b}$ is the one corresponding to $DP(h_{many}^{a,b}, r_{many}^{a,b})$.

Our global solution is $APX^* = \cup_{a,b}(APX_{many}^{a,b} \cup (\cup_{f \in F_{few}^{a,b}} APX^{f,a,b}))$, with

$$w(APX^*) = \sum_{a,b,f \in F_{few}^{a,b}} w(OPT^{f,a,b}) + \sum_{a,b} w(APX_{many}^{a,b})$$

$$\geq \sum_{a,b,f \in F_{few}^{a,b}} w(OPT^{f,a,b}) + \sum_{a,b,f \in F_{many}^{a,b}} w(R^{f,a,b})$$

$$\overset{Lem.1}{\geq} \sum_{a,b,f \in F_{few}^{a,b}} w(OPT^{f,a,b}) + \sum_{a,b,f \in F_{many}^{a,b}} (1 - O(\varepsilon))w(OPT^{f,a,b})$$

$$\geq (1 - O(\varepsilon))w(OPT_F). \qquad \qquad \square$$

We are now ready to describe the global algorithm, which is inspired by [15].

We first embed G into a longer random path G' as follows. Let $\gamma = (1/\varepsilon')^{1/\varepsilon'}$, where $1/\varepsilon' = \lceil 1/\min\{\varepsilon, 1/C\}\rceil$ (in particular, $\varepsilon' \leq \min\{\varepsilon, 1/C\}$). Let m be the number of edges in the input graph. By adding dummy edges, we can assume that $m = \gamma^\ell$ for some integer ℓ. We choose integers $x \in \{1, \ldots, m\}$ and $y \in \{1, \ldots, 1/\varepsilon'\}$ uniformly at random. Next we append x dummy edges to the left of the path and $m \cdot ((1/\varepsilon')^y - 1) - x$ dummy edges to its right. All dummy edges have capacity one. Let G' be the resulting path graph, with $m' = \gamma^\ell(1/\varepsilon')^y$ edges. We remark that this step can be easily derandomized by considering all the possible values for x and y.

We next consider the following recursive dissection of G'. We split G' into γ intervals of equal length (in terms of number of edges). Each such interval is subdivided recursively in the same way, and we halt when we reach intervals of length γ or less. We let $I_1, \ldots, I_\gamma$ denote the (direct) subintervals of interval I. We remark that intervals at level $q \geq 0$ in this dissection have length $\alpha_q := m'/\gamma^q$.

We say that a job j is at level $\ell(j)$ in this dissection if its time window W_j is fully contained in an interval $I(j)$ of level $\ell(j)$, but not of level $\ell(j) + 1$. We similarly define $\ell(i)$ and $I(i)$ for a task i. For a given interval I, let $J(I)$ be the jobs whose time window if fully contained in I, but not in any one of its subintervals. Among them, we call *good* the jobs $Gd(I)$ such that all their tasks i have $I(i) = I(j)$ (i.e., they are not fully contained in a subinterval of I), and *bad* the remaining jobs $Bd(I)$[10]. We discard from the instance all the bad jobs $Bd := \cup_I Bd(I)$.

Then we apply the following recursive algorithm, that takes as input one such interval I and a *residual capacity* u' coming from earlier calls. In the root call we use $I = G'$ and $u' = u$. Let $F(I)$ be the set of rightmost edges of the

---

[9] Intuitively, the first term in the outer max corresponds to the case that the best solution does not use job k, and the second term to the weight obtained by including some task $i \in \mathcal{B}_k$ in the solution.

[10] We call good the jobs of level ℓ by definition.

subintervals of I. We consider the twUFP instance induced by (I, u') with jobs j such that $W_j \subseteq I$ (excluding the discarded bad jobs $Bd(I)$), and we apply to this instance Lemma 2 with $F = F(I)$. We remark that the tasks T_F in this case are precisely the tasks of good jobs $Gd(I)$.

This generates a quasi-polynomial size set of pairs $\{r(I), APX(I)\}$. For each such pair $\{r(I), APX(I)\}$ the algorithm branches by solving recursively each subproblem induced by each subinterval I_i, with capacity reservation u'_i induced by $u' - r(I)$: let $APX(I_i)$ be the resulting solution. The output of this recursive call is the maximum weight solution among the solutions of type $APX(I) \cup (\cup_i APX(I_i))$. The base case is given by intervals I of length at most γ. A QPTAS for this instance is provided by Lemma 2: just choose F to be all the edges in I.

It is not hard to see that the above recursive algorithm is QPT. Furthermore, it outputs a $1 - O(\varepsilon)$ approximation of the optimal solution, restricted to the subset of good jobs $Gd := \cup_I Gd(I)$. Next lemma shows that each given job is bad with sufficiently small probability. Theorem 3 follows.

Lemma 3. *Each job is good with probability at least $1 - 3\varepsilon$.*

Proof. Let us upper bound the probability that a job j is bad. We next assume that $\ell(j) < \ell$, otherwise j is deterministically good by definition and there is nothing to show.

We say that j is *risky* if there exists q such that $\varepsilon' \alpha_q \leq t_j - s_j \leq \frac{1}{\varepsilon'} \alpha_q$. We next bound the probability that j is risky. Consider a log-scale axis and call segment the distance corresponding to a multiplicative factor of $1/\varepsilon'$. The regions of risky time-window lengths correspond to 2 segments for each value of q, separated by $1/\varepsilon' - 2$ segments which are not risky. By the random choice of y, the risky regions are shifted randomly w.r.t. the time-window lengths. Therefore each job is risky with probability at most $2/(1/\varepsilon') = 2\varepsilon' \leq 2\varepsilon$.

Let us next condition on the event that job j is not risky. Then there exists a q such that $\frac{1}{\varepsilon'} \alpha_q < t_j - s_j < \varepsilon' \alpha_{q-1}$. If $\ell(j) = q - 1$, then j is good: indeed, $\tau_j \geq (t_j - s_j)/C \geq \varepsilon'(t_j - s_j) > \alpha_q$. Thus each path P_i, $i \in \mathcal{B}_j$, is strictly longer than the level q subintervals of $I(j)$.

Due to the random choice of x, the endpoints of the intervals of level $q - 1$ are randomly shifted w.r.t the time window W_j, and $\ell(j) < q - 1$ only if W_j crosses some interval of level $q - 1$. Therefore $Pr[\ell(j) < q - 1] \leq \frac{t_j - s_j}{\alpha_{q-1}} \leq \varepsilon' \leq \varepsilon$. Altogether, job j is bad with probability at most 3ε.

Remark 1. The above QPTAS extends to the special case of bagUFP where tasks in the same bag have the same demand and weight (under the natural analogue of BTWA). In particular, it is sufficient to adapt the DP from Lemma 2. However, it does not seem to extend to the case that weights and demands are arbitrary (since in that case the same bag might influence different capacity profiles $r^{a,b}_{many}$, which therefore cannot be considered separately in the DP).

3   An Improved Approximation for bagUFP

In the *Maximum Independent Set of Rectangles* problem (MISR) we are given a collection $\mathcal{R} = \{R_1, \ldots, R_n\}$ of axis-parallel rectangles in the 2D plane, where R_i has weight w_i. Our goal is to find a maximum total weight subset of rectangles which are pairwise non-overlapping[11]. We define bagMISR as the natural generalization of MISR with bags $J = \{\mathcal{B}_1, \ldots, \mathcal{B}_h\}$.

We first present a $O(\log n / \log \log n)$ approximation for bagMISR, and then show how to use it to achieve the same approximation factor for bagUFP.

Approximating bagMISR. Let $\mathcal{P}$ be the set of $O(n^2)$ *representative* points for rectangles $\mathcal{R}$ obtained with the already mentioned construction. Consider the following natural LP relaxation for bagMISR:

$$\max \quad \sum_{R_i \in \mathcal{R}} w_i y_i \qquad\qquad (LP_{\text{bagMISR}})$$

$$\text{s.t.} \quad \sum_{R_i \in \mathcal{R}: p \in R_i} y_i \leq 1 \qquad\qquad \forall p \in \mathcal{P}$$

$$\sum_{R_i \in \mathcal{B}_j} y_i \leq 1 \qquad\qquad \forall \mathcal{B}_j \in J$$

$$y_i \geq 0 \qquad\qquad \forall R_i \in \mathcal{R}$$

The standard LP relaxation LP_{MISR} for MISR is obtained from the above LP by removing the bag constraints. Let $\mathcal{B}(R_i)$ be the set of rectangles in the same bag of R_i, and, for an arbitrary set of rectangles $\mathcal{R}' \subseteq \mathcal{R}$, let $y(\mathcal{R}') = \sum_{i \in \mathcal{R}'} y_i$. For two overlapping, distinct rectangles R_i and R_j, we say that they *corner-intersect*, and write $R_i \otimes R_j$, if one rectangle contains at least one corner of the other rectangle. Otherwise they *cross*. For an arbitrary set $\mathcal{R}' \subseteq \mathcal{R}$ and rectangle $R_i \in \mathcal{R}$, define the *resistance* η as:

$$\eta(R_i, \mathcal{R}') = \sum_{\substack{R_j \in \mathcal{R}' \setminus \mathcal{B}(R_i), \\ R_i \otimes R_j}} y_j \; + \sum_{R_j \in \mathcal{R}' \cap \mathcal{B}(R_i) \setminus \{R_i\}} y_j$$

Let G_1 and G_2 be two undirected graphs with vertex set $\mathcal{R}$ constructed as follows: if two distinct rectangles R_i and R_j are *incompatible* (i.e., they overlap or are in the same bag), then the edge (R_i, R_j) is added to G_1 if $R_i \otimes R_j$ or if they are in the same bag, otherwise the edge (R_i, R_j) is added to G_2. Of course, a subset $I \subseteq \mathcal{R}$ is a feasible solution if and only if it induces an independent set of nodes in both the graphs simultaneously.

Our approximation algorithm works as follows. First, a fractional optimum solution y of LP_{bagMISR} is found. Then a permutation Π of $\mathcal{R}$ is computed in the following manner: given the first i rectangles $\Pi_i = \{\pi_1, \ldots, \pi_i\}$ in the permutation, the $(i+1)^{th}$ element π_{i+1} (breaking ties arbitrarily) is:

$$\pi_{i+1} = \arg \min_{R_j \in \mathcal{R} \setminus \Pi_i} \eta(R_j, \mathcal{R} \setminus \Pi_i)$$

---

[11] For our goals, it is convenient to consider two rectangles as overlapping iff they overlap on a positive value area. In particular, overlapping on rectangle boundaries is allowed.

We next compute a candidate set C and an independent set $I \subseteq C$ of G_1. Initially, C and I are empty. Then, the members of the permutation are scanned in reverse order: at iteration k, the rectangle $\pi_{n-(k-1)}$ is added to C independently with probability $y(\pi_{n-(k-1)})/10$. If $\pi_{n-(k-1)}$ is added to C and $I \cup \{\pi_{n-(k-1)}\}$ is an independent set in G_1, then $\pi_{n-(k-1)}$ is also added to I.

Note that I might not be an independent set due to crossing intersections. Let Δ be the maximum clique-size of the rectangles in $G_2[I]$, that is, the maximum number of rectangle overlapping on the same point. Since the rectangles in I only have crossing intersections, $G_2[I]$ can be colored in polynomial time using Δ colors as shown in [3]. The algorithm then returns the color subclass I' of I that has the largest total weight. Clearly, I' is an independent set in both G_1 and G_2, and thus it is a feasible solution. We can bound the approximation ratio similarly to [11].

Lemma 4. *There is an expected $O(\log n / \log \log n)$ approximation for bagMISR.*

Approximating bagUFP. Our algorithm works as follows. We first compute an approximate solution APX_{small} associated to *small* tasks $T_{small} = T \setminus T_{large}$ using the algorithm in [10]. We recall that this algorithm computes a constant approximation of the optimal fractional solution of LP_{bagUFP} restricted to small tasks [10, Lemma1]. Next we focus on large tasks, and on the corresponding set of top-drawn rectangles $\mathcal{R}$. We consider the bagMISR instance induced by $\mathcal{R}$. We compute a solution $\mathcal{R}'$ for this instance using the algorithm from Lemma 4. Let APX_{large} be the tasks corresponding to $\mathcal{R}'$ (observe that APX_{large} is a feasible bagUFP solution). We finally return the best solution APX between APX_{small} and APX_{large}.

Proof (of Theorem 1). Consider the above algorithm. We prove that APX is a $O(\log n / \log \log n)$ approximation with respect to the cost opt of the optimal fractional solution x to $LP_{bagUFP+}$. Let x^{small} be the restriction of x to small tasks, and opt^{smal} be the corresponding weight. We define x^{large} and opt^{large} analogously for large tasks.

If $opt^{small} \geq opt/2$, then APX_{small} has the desired properties by [10]. Indeed, x^{small} if a feasible solution to LP_{bagUFP}.

Otherwise, let $y^{large} = x^{large}/4$. Observe that y^{large} is a feasible solution for $LP_{bagMISR}$. Therefore, APX_{large} provides a $O(\log n / \log \log n)$ approximation of the weight of y^{large} (hence of opt^{large}). The claim follows.   □

4   A $O(1)$-Approximation for Uniform Profits

In this section we present our $O(1)$ approximation for bagUFP with uniform weights, which also upper bounds the integrality gap of $LP_{bagUFP+}$ in the same case. By scaling, we can assume w.l.o.g. that weights are exactly one.

By the same argument as in the proof of Theorem 1, it is sufficient to provide a $O(1)$ approximation for the (uniform-weight) bagMISR instance induced by the top-drawn rectangles $\mathcal{R}$ corresponding to large tasks. We use as a black box the following result proved (implicitly) in [1]. We recall that in the *Maximum Independent Set of Intervals* problem (MISI) we are given a collection $\mathcal{I}$ of intervals along a line, each one with an associated weight, and our goal is to compute a maximum weight subset of intervals $\mathcal{I}'$ so that the intervals in $\mathcal{I}'$ are pairwise non-overlapping. By bagMISI we denote the natural generalization of MISI with bag constraints. We let LP_{MISI} be the standard LP for MISI, which is defined analogously to LP_{MISR}.

Lemma 5. *Let $\mathcal{R}'$ be a set of top-drawn rectangles corresponding to a subset of large tasks in an UFP instance, and let $\mathcal{R}_{max}$ be any maximal independent set of rectangles in $\mathcal{R}'$. There is a polynomial-time algorithm that computes up to 10 points $\mathcal{P}_i$ in the plane for each $R_i \in \mathcal{R}_{max}$, and four subsets $\mathcal{R}_{point}, \mathcal{R}_{top}, \mathcal{R}_{left}, \mathcal{R}_{right} \subseteq \mathcal{R}'$ that cover $\mathcal{R}'$ so that:*

1. *$\mathcal{R}_{point} = \{R_i \in \mathcal{R}' : R_i \cap \mathcal{P} \neq \emptyset\}$ with $\mathcal{P} = \cup_{R_i \in \mathcal{R}_{max}} \mathcal{P}_i$.*
2. *For each $x \in \{top, left, right\}$, there exists a bijection between $\mathcal{R}_x$ and a collection $\mathcal{I}_x$ of intervals along a line, so that the corresponding set of feasible fractional solutions to LP_{MISR} and LP_{MISI}, respectively, is the same.*

Our approximation algorithm for uniform-weight bagMISR works as follows. We compute any maximal feasible solution APX_{max} for the bagMISR instance induced by $\mathcal{R}$. Consider the rectangles $\mathcal{R}_{bag} \subseteq \mathcal{R} \setminus APX_{max}$ such that at least one (indeed, precisely one) task in the same bag is contained in APX_{max}.

We apply the algorithm from Lemma 5 with $\mathcal{R}_{max} = APX_{max}$ and $\mathcal{R}' = \mathcal{R} \setminus \mathcal{R}_{bag}$[12]. This way we obtain the sets $\mathcal{R}_{point}, \mathcal{R}_{top}, \mathcal{R}_{left}, \mathcal{R}_{right}$. For any $x \in \{top, left, right\}$, we consider the instance of (uniform weight) bagMISI induced by $\mathcal{I}_x$ (where the bags are defined by the corresponding bijection). We apply the LP-based 2-approximation algorithm for bagMISI in [6] to this instance, hence obtaining an approximate solution APX_x. Finally, we output the best solution APX among the solutions APX_{max}, APX_{top}, APX_{left}, and APX_{right}.

Lemma 6. *The above algorithm is a 17 approximation for the bagMISR instances induced by large tasks of a bagUFP instance.*

Proof. Let y be the optimal fractional solution to $LP_{bagMISR}$ with weight opt. For $x \in \{top, left, right, bag, point\}$, let y_x be the restriction of y to rectangles $\mathcal{R}_x$, and let opt_x be the corresponding fractional weight.

Suppose that $opt_x \geq 2opt/17$ for some $x \in \{top, left, right\}$. Since y_x is feasible for $LP_{bagMISI}$ on intervals $\mathcal{I}_x$, then $|APX_x| \geq opt_x/2 \geq opt/17$.

Suppose next that $opt_{bag} \geq opt/17$. Let $\mathcal{B}_j$ be a bag corresponding to some task $i \in APX_{max}$. The total weight in y_{bag} for this bag is at most 1 by the bag constraints. Therefore $|APX_{max}| \geq opt_{bag} \geq opt/17$.

Finally, assume $opt_{point} \geq 10opt/17$. Next let $\mathcal{R}_p \subseteq \mathcal{R}_{point}$ be the rectangles containing some point $p \in \mathcal{P}_i$ for some rectangle $R_i \in APX_{max}$. The optimal

---

[12] Observe that, by construction, APX_{max} is a maximal independent set w.r.t $\mathcal{R}'$. This might not be the case w.r.t $\mathcal{R}$ since bag constraints might prevent some non-overlapping rectangle to be included in the maximal solution.

fractional weight associated to $\mathcal{R}_p$ is at most 1 due to the LP constraints. Therefore, $opt_{point} \leq \sum_{R_i \in APX_{max}} \sum_{p \in \mathcal{P}_i} 1 \leq 10 \sum_{R_i \in APX_{max}} 1 = 10|APX_{max}|$. Thus $|APX_{max}| \geq opt_{point}/10 \geq opt/17$. □

Theorem 2 follows from Lemma 6 and the above discussion.

Acknowledgements. The authors wish to thank Andreas Wiese for very helpful discussions about UFP and related problems.

References

1. Anagnostopoulos, A., Grandoni, F., Leonardi, S., Wiese, A.: Constant integrality gap LP formulations of unsplittable flow on a path. In: Goemans, M., Correa, J. (eds.) IPCO 2013. LNCS, vol. 7801, pp. 25–36. Springer, Heidelberg (2013)
2. Anagnostopoulos, A., Grandoni, F., Leonardi, S., Wiese, A.: A mazing 2+ε approximation for unsplittable flow on a path. In: SODA, pp. 26–41 (2014)
3. Asplund, E., Grünbaum, B.: On a coloring problem. Math. Scand **8**, 181–188 (1960)
4. Bansal, N., Chakrabarti, A. Epstein,, A., Schieber, B.: A quasi-PTAS for unsplittable flow on line graphs. In: STOC, pp. 721–729 (2006)
5. Bansal, N., Friggstad, Z., Khandekar, R., Salavatipour, R.: A logarithmic approximation for unsplittable flow on line graphs. In: SODA, pp. 702–709 (2009)
6. Bar-Noy, A., Guha, S., Naor, J., Schieber, B.: Approximating the throughput of multiple machines in real-time scheduling. SIAM J. Comput. **31**(2), 331–352 (2001)
7. Batra, J., Garg, N., Kumar, A., Mömke, T., Wiese, A.: New approximation schemes for unsplittable flow on a path. In: SODA, pp. 47–58 (2015)
8. Bonsma, P., Schulz, J., Wiese, A.: A constant factor approximation algorithm for unsplittable flow on paths. In: FOCS, pp. 47–56 (2011)
9. Calinescu, G., Chakrabarti, A., Karloff, H., Rabani, Y.: Improved approximation algorithms for resource allocation. In: Cook, W.J., Schulz, A.S. (eds.) IPCO 2002. LNCS, vol. 2337, pp. 401–414. Springer, Heidelberg (2002)
10. Chakaravarthy, V.T., Choudhury, A.R., Gupta, S., Roy, S., Sabharwal, Y.: Improved algorithms for resource allocation under varying capacity. In: Schulz, A.S., Wagner, D. (eds.) ESA 2014. LNCS, vol. 8737, pp. 222–234. Springer, Heidelberg (2014)
11. Chan, T.M., Har-Peled, S.: Approximation algorithms for maximum independent set of pseudo-disks. Discrete Comput. Geom. **48**(2), 373–392 (2012)
12. Chekuri, C., Ene, A., Korula, N.: Unsplittable flow in paths and trees and column-restricted packing integer programs. In: Dinur, I., Jansen, K., Naor, J., Rolim, J. (eds.) Approximation, Randomization, and Combinatorial Optimization. Algorithms and Techniques. LNCS, vol. 5687, pp. 42–55. Springer, Heidelberg (2009)
13. Chekuri, C., Mydlarz, M., Shepherd, F.: Multicommodity demand flow in a tree and packing integer programs. ACM Trans. Algorithms **3**, 27 (2007)
14. Darmann, A., Pferschy, U., Schauer, J.: Resource allocation with time intervals. Theor. Comput. Sci. **411**, 4217–4234 (2010)
15. Grandoni, F., Rothvoß, T.: Pricing on paths: a PTAS for the highway problem. In: SODA, pp. 675–684 (2011)
16. Spieksma, F.: On the approximability of an interval scheduling problem. J. Sched. **2**(5), 215–227 (1999)

Maximum ATSP with Weights Zero
and One via Half-Edges

Katarzyna Paluch[(⊠)]

Wroclaw University, Wroclaw, Poland
abraka@cs.uni.wroc.pl

Abstract. We present a fast combinatorial 3/4-approximation algorithm for the maximum asymmetric TSP with weights zero and one. The approximation factor of this algorithm matches the currently best one given by Bläser in 2004 and based on linear programming. Our algorithm first computes a maximum size matching and a maximum weight cycle cover without certain cycles of length two but possibly with *half-edges* - a half-edge of a given edge e is informally speaking a half of e that contains one of the endpoints of e. Then from the computed matching and cycle cover it extracts a set of paths, whose weight is large enough to be able to construct a traveling salesman tour with the claimed guarantee.

1 Introduction

We study the maximum asymmetric traveling salesman problem with weights zero and one (Max (0,1)-ATSP), which is defined as follows. Given a complete loopless directed graph G with edge weights zero and one, we wish to compute a traveling salesman tour of maximum weight. Traveling salesman problems with weights one and two are an important special case of traveling salesman problems with triangle inequality. Max (0,1)-ATSP is connected to Min (1,2)-ATSP (the minimum asymmetric traveling salesman problem with weights one and two) in the following way. It has been shown by Vishvanathan [17] that a $(1 - \alpha)$-approximation algorithm for Max (0,1)-ATSP yields a $(1 + \alpha)$-approximation algorithm for Min (1,2)-ATSP by replacing weight two with weight zero.

Approximating Max (0,1)-ATSP with the ratio 1/2 is easy – it suffices to compute a maximum weight matching of the graph G and patch the edges arbitrarily into a tour. The first nontrivial approximation of Max (0,1)-ATSP was given by Vishvanathan [17] and has the approximation factor 7/12. It was improved on by Kosaraju, Park, and Stein [8] in 1994, who gave a 48/63-approximation algorithm that also worked for Max ATSP with arbitrary nonnegative weights. Later, Bläser and Siebert [4] obtained a 4/3-approximation algorithm for Min (1,2)-ATSP, which can be modified to give a 2/3-approximation algorithm for Max (0,1)-ATSP. 2/3-approximation algorithms are also known for the general Max ATSP and have been given in [6] and [14]. The currently best published

Partly supported by Polish National Science Center grant UMO-2013/11/B/ST6/01748.

L. Sanità and M. Skutella (Eds.): WAOA 2015, LNCS 9499, pp. 25–34, 2015.
DOI: 10.1007/978-3-319-28684-6_3

approximation algorithm for Max (0,1)-ATSP achieving ratio 3/4 is due to Bläser [2]. It uses linear programming to obtain a multigraph G_M of weight at least 3/2 times the weight of an optimal traveling salesman tour (OPT) such that G_M can be *path-2-colored*. A multigraph is called *path-2-colorable* if its edges can be colored with two colors so that each color class consists of vertex-disjoint paths. The algorithm by Bläser has a polynomial running time but the degree of the polynomial is high. A 3/4-approximation algorithm for Max ATSP with arbitrary nonnegative weights has been given in [15]. The algorithm presented here for Max (0,1)-ATSP is much simpler than the one in [15].

Karpinski and Schmied have shown in [7] that it is NP-hard to approximate Min (1,2)-ATSP with an approximation factor less than 207/206 and for the general Max ATSP that it is NP-hard to obtain an approximation better than 203/204.

Our Approach and Results. We present a simple combinatorial 3/4-approximation algorithm for Max (0,1)-ATSP. First we compute a maximum weight matching M_{max} of G. By a matching of G we mean any vertex-disjoint collection of edges. The weight of M_{max} is clearly at least OPT/2, where OPT denotes the weight on an optimal tour. Next, we compute a maximum weight *cycle cover* that *evades the matching* M_{max}. A *cycle cover* of a directed graph is a collection of directed cycles such that each vertex belongs to exactly one cycle of the collection. A *cycle cover of a graph G that evades a matching M* is a cycle cover of G which does not contain any length two cycle (called a *2-cycle*) going through two vertices that are connected by some edge of M but it may contain *half-edges* - a half-edge of a given edge e is informally speaking a half of e that contains one of the endpoints of e. Half-edges have already been introduced in [14]. The task of finding a maximum weight cycle cover C_{max} that evades a matching M can be reduced to finding a maximum size matching in an appropriately constructed graph. The weight of C_{max} is an upper bound on OPT. Further on we show that a maximum weight matching M_{max} and a maximum weight cycle cover that evades M_{max} can be easily transformed into a path-2-colorable multigraph. For completeness we give also our own linear time procedure of path-2-coloring. This method takes advantage of the fact that the edge weights are zero and one. A more general algorithm for path-2-coloring that runs in $O(n^3)$ has been given in [2].

This way the main results of this paper can be stated as

Theorem 1. *There exists a combinatorial 3/4-approximation algorithm for Max (0,1)-ATSP. Its running time is $O(n^{1/2}m)$, where n and m denote the number of respectively vertices and edges of weight one in the graph.*

Corollary 1. *There exists a combinatorial 5/4-approximation algorithm for Min (1,2)-ATSP. Its running time is $O(n^{1/2}m)$.*

2   Cycle Cover that Evades Matching M

The algorithm for Max (0,1)-ATSP starts from computing a maximum weight perfect matching M_{max} of G. By a *0-edge* and a *1-edge* we will mean an edge of weight, respectively, zero or one. By G_1 we denote the subgraph of G consisting of all 1-edges of G. In order to obtain a maximum weight perfect matching M_{max} of G, it is enough to compute a maximum size matching M_1 in G_1 and, if necessary, complete it arbitrarily with 0-edges so that the resulting matching is perfect.

Next, we would like to find a maximum weight cycle cover of G that does not contain any 2-cycle in G_1, whose one edge belongs to M_{max}. Since computing such a cycle cover is NP-hard, which follows from a similar result proved in [4], we are going to relax the notion of a cycle cover and allow it to contain **half-edges** - a half-edge of edge (u, v) is informally speaking "half of the edge (u, v) that contains either a head or a tail of (u, v)".

Now, we are going to give a precise definition of a cycle cover that evades a matching M. We say that a 2-cycle c in G_1 is M-**hit** if one of the edges of c belongs to M. We introduce a graph $\tilde{G}_M$. $\tilde{G}_M = (\tilde{V}, \tilde{E})$ is the graph obtained from G by splitting each edge (u, v) belonging to an M-hit 2-cycle of G_1 with a vertex $x_{(u,v)}$ into two edges $(u, x_{(u,v)})$ and $(x_{(u,v)}, v)$, each with weight $\frac{1}{2}w(u, v)$, where $w(u, v)$ denotes the weight of the edge (u, v). Each of the edges $(u, x_{(u,v)}), (x_{(u,v)}, v)$ is called **a half-edge (of** $(u, v))$. For any subset of edges $E' \subseteq E$ by $w(E')$ we mean $\sum_{e \in E'} w(e)$.

Definition 1. *A **cycle cover that evades a matching** M is a subset $\tilde{C} \subseteq \tilde{E}$ such that*

   (i) each vertex in V has exactly one outgoing and one incoming edge in $\tilde{C}$;
   (ii) for each M-hit 2-cycle of G_1 connecting vertices u and v $\tilde{C}$ contains either zero or two edges from
      $\{(u, x_{(u,v)}), (x_{(u,v)}, v), (v, x_{(v,u)}), (x_{(v,u)}, u)\}$. *Moreover, if $\tilde{C}$ contains only one half-edge of (u, v), then it also contains one half-edge of (v, u), and one of these half-edges is incident with u and the other with v.*

To compute a cycle cover C_1 that evades M_{max} we construct the following undirected graph $G' = (V', E')$. For each vertex v of G we add two vertices v_{in}, v_{out} to V'. For each edge $(u, v) \in E$ we add vertices e^1_{uv}, e^2_{uv}, an edge (e^1_{uv}, e^2_{uv}) of weight 0 and edges $(u_{out}, e^1_{uv}), (v_{in}, e^2_{uv})$, each of weight $\frac{1}{2}w(u, v)$. Next we build so-called gadgets.

For each M-hit 2-cycle in G_1 on vertices u and v we add vertices $a_{\{u,v\}}, b_{\{u,v\}}$ and edges $(a_{\{u,v\}}, e^1_{uv}), (a_{\{u,v\}}, e^2_{vu}), (b_{\{u,v\}}, e^1_{vu}), (b_{\{u,v\}}, e^2_{uv})$ having weight 0.

Theorem 2. *Any perfect matching of G' yields a cycle cover C_1 that evades M_{max}. A maximum weight perfect matching of G' yields a cycle cover C_{max} that evades M_{max} such that $w(C_{max}) \geq OPT$.*

Proof. The proof of the first statement is very similar to the proof of Lemma 2 in [14]. We include it here for completeness. Suppose that a 2-cycle in G_1 on vertices u and v is M_{max}-hit. Then in G' there exists a gadget with vertices $a_{\{u,v\}}$ and $b_{\{u,v\}}$. In a perfect matching of G' vertex $a_{\{u,v\}}$ can be matched only with e^1_{uv} or e^2_{vu}. Similarly, vertex $b_{\{u,v\}}$ can be matched only with e^2_{uv} or e^1_{vu}.

Let us consider the case when $a_{\{u,v\}}$ is matched with e^1_{uv} and $b_{\{u,v\}}$ is matched with e^2_{uv}. Then either e^2_{vu} is matched with e^1_{vu} or e^2_{vu} is matched with u_{in} and e^1_{vu} is matched with v_{out}. The first of these scenarios means that C_1 does not contain any half-edge of (u, v) or any half-edge of (v, u). The second of these scenarios means that C_1 contains a whole edge (v, u) (both of its half-edges) and none of the half-edges of (u, v).

Suppose now that in a perfect matching of G' vertex $a_{\{u,v\}}$ is matched with e^1_{uv} and $b_{\{u,v\}}$ is matched with e^1_{vu}. Then e^2_{uv} must be matched with v_{in} and e^2_{vu} with u_{in}. This means that C_1 contains one half-edge of (u, v) (the one incident with v) and one half-edge of (v, u) (incident with u). This way we have shown that C_1 satisfies property (ii) of Definition 1. Property (i) is also satisfied because a perfect matching of G' matches each vertex v_{in} and v_{out}.

The second statement of the lemma follows from the fact that a traveling salesman tour is also a cycle cover that evades M_{max}. □

In the following by a half-edge of a cycle cover C we will mean such a half-edge of a certain edge e contained in C that C contains only one half-edge of e. A cycle cover that evades a matching M consists of directed cycles and/or directed paths, where each of the directed paths begins and ends with a half-edge. From a matching M_{max} and a cycle cover C_{max} that evades M_{max} we build a multigraph G_m as follows. Basically G_m consists of one copy of M_{max} and one copy of C_{max}. However, we do not want G_m to contain half-edges. Therefore we modify C_{max} by replacing each pair of half-edges of edges connecting vertices u and v that are contained in C_{max} with an edge (u, v), if M_{max} contains (v, u) and otherwise with an edge (v, u). As a result G_m contains a 2-cycle on each such pair of vertices u, v. After this modification C_{max} contains only whole edges and may contain directed paths with a common endpoint i.e., some vertices may have indegree two and outdegree zero or vice versa. However, the overall weight of C_{max} is unchanged. Now, G_m is going to contain two copies of an edge e if e belongs both to M_{max} and C_{max} and one copy of an edge e if e belongs either to M_{max} or to C_{max}. This way we obtain a multigraph that satisfies the following conditions:

– each vertex in G_m has degree three,
– each vertex in G_m has indegree at most two and outdegree at most two,
– for each pair of vertices u and v, G_m contains at most two edges connecting u and v.

In [2] Bläser shows how to slightly modify such a multigraph so that it has the same number of 1-edges and is path-2-colorable. Path-2-coloring of the modified graph is based on a variant of the path-2-coloring lemma given by Lewenstein and Sviridenko [11], which in turn is a reduction to the path-2-coloring lemma of Kosaraju, Park, and Stein, whose proof was given in [1]. The running time of the path-2-coloring algorithm is $O(n^3)$.

If the number of vertices in the graph is odd, then the above approach does not give a 3/4-approximation. We can either add a new additional vertex, that is connected to every other vertex by a 0-edge and obtain a $3/4(1 - 1/n)$-approximation, or guess one edge of an optimal traveling salesman tour and contract it. In the latter case, the running time of the algorithm becomes $O(n^{3/2}m)$.

3   Path-2-coloring

In this section we present a simple linear time algorithm of path-2-coloring the multigraph G_m computed in the previous section.

From G_m we are going to obtain another multigraph that contains the same number of 1-edges as G_m and additionally allows a simple method of path-2-coloring.

3.1   Eliminating 2-cycles

First we deal with 2-cycles on cycles and paths of C_{max}. For any 1-edge $e = (u, v)$ contained in a cycle c of C_{max} such that M_{max} contains a 1-edge $e' = (v, u)$, we replace the edge e' with another copy of e. Similarly, for any 1-edge $e = (u, v)$ contained in a path p of C_{max} such that e has not been obtained from a half-edge and M_{max} contains a 1-edge $e' = (v, u)$, we replace the edge e' with another copy of e. So far, clearly, we have not diminished the number of 1-edges contained in G_m. Next, we discard all 0-edges from G_m. This way, some cycles of C_{max} disintegrate into paths and some paths of C_{max} give rise to shorter paths. In what follows, by a cycle of C_{max} we will mean a cycle of C_{max} consisting solely of 1-edges and by a path of C_{max} we will mean a maximal (under inclusion) directed path, whose every edge belongs to C_{max} and has weight one.

Let $e = (u, v)$ be an edge of G_1, c a cycle and p a path of C_{max}. Then we say that e is an ***inray of*** c ***(corr. p)*** if $u \notin c$ and $v \in c$ (corr. $u \notin p$ and $v \in p$). If $u \in c$ and $v \notin c$ (corr. $u \in p$ and $v \notin p$), then we say that e is an ***outray of*** c ***(corr. p)***. A ***ray of*** c (p) is any inray or outray of c (p). If both endpoints of e belong to c (corr. p) and e does not belong to c (corr. p), then e is called a ***chord*** of c (corr. p). If e is a copy of some edge belonging to c (corr. p), then e is called an ***ichord***.

Let us notice that any 2-cycle which is present at this stage of G_m is either a 2-cycle of C_{max} or a 2-cycle obtained from a pair of half-edges of C_{max} and an edge of M_{max}. Now, we construct from G_m a new multigraph G_m^1 by eliminating all remaining 2-cycles as follows. If c is a 2-cycle of C_{max} on vertices u and v that has an inray incident to u and an outray incident to v, then in G_m we replace the edge (v, u) with another copy of edge (u, v) and in G_m^1 we shrink the two copies of an edge (u, v) into a single vertex. By shrinking an edge (u, v) into a vertex we mean removing an edge (u, v), replacing vertices u and v with one new vertex and replacing any edge of G_m incident to u or v with an edge incident to the newly added vertex. Every remaining 2-cycle of C_{max} or a 2-cycle obtained from a pair of half-edges of C_{max} and an edge of M_{max} is also shrunk into a single vertex in G_m^1. We continue such shrinking until G_m^1 contains no 2-cycles.

We make the following observation.

Observation 1. *From any path-2-coloring of G_m^1 we can obtain a path-2-coloring of G_m without changing the color of any edge of G_m^1.*

3.2   Flipping of Edges

Next we are going to further flip some of the edges of G_m^1 and obtain a graph G_m^2 such that the task of its path-2-coloring is very easy.

For each cycle c of C_{max} we are going to flip either its inrays and chords or outrays and chords so that c has either only outrays and ichords or only inrays and ichords. Let c be any cycle of C_{max}. Let us notice that its length is at least three. Suppose that the number of inrays of c is not smaller than the number of outrays of c. Then we replace the outrays and chords of c with ichords of c in such a way that the indegree and outdegree of each vertex of c is at most two. More precisely, the replacement is carried out as follows. Let F_c be a set of *free edges* of c, where we say that an edge (u, v) of c is free if no inray of c is incident with v. The number of free edges of c is not smaller than the number of outrays and chords of c. Moreover, the number of outrays and chords of c is not bigger than $|c| - 2$, where $|c|$ denotes the length of c. It follows from the fact that the number of chords of c is not greater than $|c|/2$ and the number of outrays of c does not exceed the number of inrays of c. Each chord and outray of c is then replaced with a copy of some distinct edge of F_c. Let us notice that it may happen that as a result of this operation some vertex of c has both indegree and outdegree equal to two.

If the number of outrays of c outnumbers the number of inrays of c, then we replace the inrays and chords of c with ichords of c in an analogous way as above.

fact 1. *Let c be any cycle that has either only inrays and/or ichords or only outrays and/or ichords. Moreover, (1) the number of rays of c is at least two or c has at most $|c| - 2$ ichords and (2) the indegree and outdegree of each vertex of c is at most two. Assume also that each ray of c has already been colored with 1 or 2. If c has at least two rays, then it is possible to path-2-color the edges and ichords of c provided that two rays of c are colored differently. If c has at most one ray, then it is always possible to path-2-color the edges and ichords of c.*

Proof. Any two copies of the same edge must be colored differently. Similarly any two outgoing edges of some vertex of c or any two incoming edges of some vertex of c must be colored differently. If c has two rays that are colored differently, then two edges of c incident to these rays must also be colored differently and it follows that no monochromatic cycle can arise out of the edges or ichords of c. If c has exactly one ray colored with, say 1, then we must see to it that not for every edge (u, v) of c it is that at least one copy of (u, v) is colored with 2. Since c has at most $|c| - 2$ ichords, there exists an edge e of c such that G_m^2 contains only one copy of e and which can be colored with 1. If c has no rays, then we can easily path-2-color its edges and ichords.   □

The situation with paths is slightly more complicated. We are going to distinguish paths that are **bound** and **free**. A path of C_{max} is said to be bound if it shares at least one of its endpoints with another path of C_{max}. A path of C_{max} that is not bound is said to be free. A bound path can be **1-bound** – if exactly one of its endpoints is also an endpoint of another path of C_{max} or **2-bound** – if each of its endpoints is an endpoint of another path of C_{max}. We say that an edge $e = (u, v)$ of p of C_{max} is a **rayter** if u is incident with an outray of p and v is incident with an inray of p.

We are going to flip the rays and chords of each bound path p in such a way that besides possible ichords p either has at most one ray or exactly two rays incident to a rayter. As for free paths we are going to flip the rays and chords of each free path p in such a way that besides possible ichords p either has only inrays or only outrays or exactly two rays incident to a rayter.

Let p be any path of C_{max} with endpoints u and v. By $|p|$ we denote the length of p i.e., the number of edges of p. An endpoint of p which is not an endpoint of any other path of C_{max} is said to be a **border vertex** of p. If an endpoint u of p belongs also to some other path of C_{max}, then the edge of p incident to u is called a **border edge** of p. The endpoint of a border edge of p that is not an endpoint of any path of C_{max} different from p is also called a **border vertex** of p. It may happen that a path p of C_{max} does not have any border vertex – if $|p| = 1$ and both endpoints of p belong also to some other path(s) of C_{max}. We say that a path p has a **good ray** if it has a ray e incident to a border vertex v of p such that either (1) v is an endpoint of p and e together with p form a directed path of length $|p| + 1$ or (2) v is not an endpoint of p and e forms a directed path of length two with e', where e' is an edge of p incident to v and is not a border edge of p. For example, let p be a 2-bound path (u, v_1, v_2, v) directed from u to v and suppose that p has a ray $e = (v_2, v_3)$. Then e is a good ray of p. Let us notice that the maximum number of edges of M_{max} incident to a path p of C_{max} is: (1) $|p| - 1$, if p is 2-bound, (2) $|p|$, if p is 1-bound and (3) $|p| + 1$, if p is free. It is so because no edge of M_{max} is incident to a vertex which is an endpoint of two different paths of C_{max} – because such an endpoint is in fact a shrunk 2-cycle.

The flipping of rays and ichords of paths proceeds as follows. If the number of edges of M_{max} incident to a given path p is (1) fewer than $|p| - 1$ and p is 2-bound or (2) fewer than $|p|$ and p is 1-bound or (3) fewer than $|p| + 1$ and p is free, then we replace all chords and rays of p with ichords of p but not with a copy of any border edge of p. (Also, of course, no edge of p is allowed to occur in more than two copies). Otherwise, if a path p has a good ray, we keep any one good ray of p and replace all the other rays and chords of p with ichords of p but not with a copy of any border edge of p. In the remaining case, we keep some two rays of p that are incident to a rayter and replace the rest of rays and chords of p with ichords.

fact 2. *Graphs G_m^1 and G_2^m are on the same set of vertices and have the same number of edges.*

3.3   Path-2-coloring of G_m^2

Suppose that e_1 and e_2 are good rays of paths p_1, p_2 having a common endpoint u such that both e_1 and e_2 is incident to the border edge (of respectively p_1 or p_2) incident with u. Then the rays e_1 and e_2 are said to be **allied**.

We make the following two observations.

fact 3. *In any path-2-coloring of G_m^2 the rays incident to the same rayter are colored with the same color.*

Proof. Let $e = (u, v)$ be a rayter of p. Then in any path-2-coloring of G_m^2 the edge e must be colored with a different color than an outray of p incident to u and also with a different color than an inray of p incident to v. Since there are only two colors, it follows that the rays incident to e must be colored with the same color.   □

fact 4. *In any path-2-coloring of G_m^2 the allied rays are colored with different colors.*

Proof. Let v be a vertex which is an endpoint of two different paths p_1, p_2 of C_{max} and let e_1, e_2 be two border edges incident to v. Then, clearly e_1 and e_2 must be colored with different colors as either both are incoming edges of v or both are outgoing edges of v. The ray incident to e_1 must be colored differently than e_1. Similarly the ray incident to e_2 must be colored differently than e_2.   □

After all the flipping, the multigraph G_m^2 is quite easy to path-2-color. In fact, it suffices to appropriately color the rays and then the coloring of the rest of the edges is straightforward. From the rays in G_m^2 we build the following graph H. At the beginning H has the same vertex set as G_m^2 and contains all the rays in G_m^2, i.e., (u, v) is an edge in H if and only if (u, v) is a ray of some path or cycle of C_{max} in G_m^2. Thus every two edges of H are vertex-disjoint at this stage. Next, for each cycle c of C_{max} we choose two arbitrary rays e_1, e_2 of c and merge together their endpoints belonging to c i.e., if $u_1 \in e_1 \cap c$ and $u_2 \in e_2 \cap c$, then we replace u_1 and u_2 with one vertex and as a result e_1 and e_2 have (at least) one common endpoint. Further, each pair of rays incident to the same rayter is replaced with one edge as follows. Let $e_1 = (u_1, v_1), e_2 = (u_2, v_2)$ be a pair of rays incident to some edge $e = (u_2, v_1)$ in G_m^2. Then e_1, e_2 are replaced in H with one edge $e = (u_1, v_2)$. Such replacements are done exhaustively. We also merge together the endpoints of certain pairs of good rays. Suppose that e_1 and e_2 are allied rays of paths p_1, p_2. Then we merge together the endpoint of e_1 belonging to p_1 with the endpoint of e_2 belonging to p_2.

At this stage, ignoring the directions H consists of paths, cycles and isolated vertices i.e., each vertex is either isolated or belongs to exactly one path or cycle. Moreover, every two edges of H sharing a vertex v are either both incoming edges of v or are both ougoing-edges of v. Hence each cycle of H has even length. We color the edges of each cycle and each path of H with 1 and 2 in such a way that no two incoming edges of any vertex are colored with the same color or no two outgoing edge of any vertex are colored with the same color. In other words, we path-2-color H.

Lemma 1. *Any path-2-coloring of H can be extended to a path-2-coloring of G_m^2.*

Proof. Each ray in G_m^2 is colored with the same color as in H. In the case when some edge e in H was obtained from several rays in G_m^2, each such ray in G_m^2 is colored in the same way as e in H. Thus, by the way we constructed H, each pair of rays incident to one rayter is colored in the same way, allied rays are colored with different colors and for each cycle c of C_{max} that has at least two rays, there exist two rays of c colored differently. By Fact 1 we already know how to color the edges and ichords of each cycle of C_{max}. Any edge $e = (u, v)$ of any path of C_{max} which is incident to an outray r_1 incident to u is colored differently than r_1. Similarly any edge $e = (u, v)$ of any path of C_{max} which is incident to an inray r_2 incident to v is colored differently than r_2. Also two border edges of two different paths of C_{max} incident to the same vertex are colored differently. Two copies of the same edge are clearly colored differently. The remaining edges can be colored arbitrarily. ☐

Observation 2. *From any path-2-coloring of G_m^2 we can obtain a path-2-coloring of G_m without changing the color of any edge of G_m^2.*

References

1. Bläser, M.: An 8/13-approximation algorithm for the asymmetric maximum TSP. J. Algorithms **50**(1), 23–48 (2004)
2. Bläser, M.: A 3/4-approximation algorithm for maximum ATSP with weights zero and one. In: Jansen, K., Khanna, S., Rolim, J.D.P., Ron, D. (eds.) RANDOM 2004 and APPROX 2004. LNCS, vol. 3122, pp. 61–71. Springer, Heidelberg (2004)
3. Bläser, M., Manthey, B.: Two approximation algorithms for 3-cycle covers. In: Jansen, K., Leonardi, S., Vazirani, V.V. (eds.) APPROX 2002. LNCS, vol. 2462, p. 40. Springer, Heidelberg (2002)
4. Bläser, M., Siebert, B.: Computing cycle covers without short cycles. In: Meyer auf der Heide, F. (ed.) ESA 2001. LNCS, vol. 2161, pp. 369–379. Springer, Heidelberg (2001)
5. Fisher, M.L., Nemhauser, G.L., Wolsey, L.A.: An analysis of approximations for finding a maximum weight Hamiltonian circuit. Oper. Res. **27**(4), 799–809 (1979)
6. Kaplan, H., Lewenstein, M., Shafrir, N., Sviridenko, M.: Approximation algorithms for asymmetric tsp by decomposing directed regular multigraphs. J. ACM **52**(4), 602–626 (2005). Preliminary version appeared in FOCS'03
7. Karpinski, M., Schmied, R.: Improved Inapproximability results for the shortest superstring and related problems. In: CATS, pp. 27–36 (2013)
8. Kosaraju, S.R., Park, J.K., Stein, C.: Long tours and short superstrings (preliminary version). In: Proceedings of the 35th Annual Symposium on Foundations of Computer Science, pp. 166–177 (1994)

9. Kowalik, L., Mucha, M.: Deterministic 7/8-approximation for the metric maximum tsp. Theor. Comput. Sci. **410**(47–49), 5000–5009 (2009)
10. Kowalik, L., Mucha, M.: 35/44-approximation for asymmetric maximum tsp with triangle inequality. Algorithmica **59**(2), 240–255 (2011)
11. Lewenstein, M., Sviridenko, M.: A 5/8 approximation algorithm for the maximum asymmetric tsp. SIAM J. Discrete Math. **17**(2), 237–248 (2003)
12. Lovasz, L., Plummer, M.D.: Matching Theory (1986)
13. Paluch, K., Mucha, M., Madry, A.: A 7/9 - approximation algorithm for the maximum traveling salesman problem. In: Dinur, I., Jansen, K., Naor, J., Rolim, J. (eds.) Approximation, Randomization, and Combinatorial Optimization. LNCS, vol. 5687, pp. 298–311. Springer, Heidelberg (2009)
14. Paluch, K.E., Elbassioni, K.M., van Zuylen, A.: Simpler approximation of the maximum asymmetric traveling salesman problem. In: Proceedings of the 29th Symposium on Theoretical Aspects of Computer Science, STACS 2012, Leibniz International Proceedings of Informatics 14, pp. 501–506 (2012)
15. Paluch, K.: Better Approximation Algorithms for Maximum Asymmetric Traveling Salesman and Shortest Superstring. CoRR abs/1401.3670 (2014)
16. Papadimitriou, C.H., Yannakakis, M.: The traveling salesman problem with distances one and two. Math. Oper. Res. **18**, 1–11 (1993)
17. Vishwanathan, S.: An approximation algorithm for the asymmetric travelling salesman problem with distances one and two. Inform. Proc. Lett. **44**, 297–302 (1992)

An FPT 2-Approximation for Tree-cut Decomposition

Eunjung Kim[1], Sang-il Oum[2], Christophe Paul[3],
Ignasi Sau[3], and Dimitrios M. Thilikos[3,4,5(✉)]

[1] CNRS, LAMSADE, Paris, France
eunjungkim78@gmail.com
[2] Department of Mathematical Sciences, KAIST, Daejeon, South Korea
sangil@kaist.edu
[3] CNRS, Université de Montpellier, LIRMM, Montpellier, France
{christophe.paul,ignasi.sau}@lirmm.fr,
sedthilk@thilikos.info
[4] Department of Mathematics, University of Athens, Athens, Greece
[5] Computer Technology Institute Press "Diophantus", Patras, Greece

Abstract. The tree-cut width of a graph is a graph parameter defined by Wollan [*J. Comb. Theory, Ser. B, 110:47–66, 2015*] with the help of tree-cut decompositions. In certain cases, tree-cut width appears to be more adequate than treewidth as an invariant that, when bounded, can accelerate the resolution of intractable problems. While designing algorithms for problems with bounded tree-cut width, it is important to have a parametrically tractable way to compute the exact value of this parameter or, at least, some constant approximation of it. In this paper we give a parameterized 2-approximation algorithm for the computation of tree-cut width; for an input n-vertex graph G and an integer w, our algorithm either confirms that the tree-cut width of G is more than w or returns a tree-cut decomposition of G certifying that its tree-cut width is at most $2w$, in time $2^{O(w^2 \log w)} \cdot n^2$. Prior to this work, no *constructive* parameterized algorithms, even approximated ones, existed for computing the tree-cut width of a graph. As a consequence of the Graph Minors series by Robertson and Seymour, only the *existence* of a decision algorithm was known.

Keywords: Fixed-parameter tractable algorithm · Tree-cut width · Approximation algorithm

The research of the last author was co-financed by the European Union (European Social Fund ESF) and Greek national funds through the Operational Program "Education and Lifelong Learning" of the National Strategic Reference Framework (NSRF), Research Funding Program: ARISTEIA II. The second author was supported by Basic Science Research Program through the National Research Foundation of Korea (NRF) funded by the Ministry of Science, ICT & Future Planning (2011-0011653).

A complete version of this extended abstract has appeared in arXiv.org as http://arxiv.org/abs/1509.04880. The proofs of the results marked with '(*)' can be found there.

© Springer International Publishing Switzerland 2015
L. Sanità and M. Skutella (Eds.): WAOA 2015, LNCS 9499, pp. 35–46, 2015.
DOI: 10.1007/978-3-319-28684-6_4

1   Introduction

One of the most popular ways to decompose a graph into smaller pieces is given by the notion of a tree decomposition. Intuitively, a graph G has a tree decomposition of small width if it can be decomposed into small (possibly overlapping) pieces that are altogether arranged in a tree-like structure. The *width* of such a decomposition is defined as the minimum size of these pieces. The graph invariant of *treewidth* corresponds to the minimum width of all possible tree decompositions and, that way, serves as a measure of the topological resemblance of a graph to the structure of a tree. The importance of tree decompositions and treewidth in graph algorithms resides in the fact that a wide family of NP-hard graph problems admits FPT-algorithms, i.e., algorithms that run in $f(w) \cdot n^{O(1)}$ steps, when parameterized by the treewidth w of their input graph. According to the celebrated theorem of Courcelle, for every problem that can be expressed in Monadic Second Order Logic (MSOL) [5] it is possible to design an $f(w) \cdot n$-step algorithm on graphs of treewidth at most w. Moreover, towards improving the parametric dependence, i.e., the function f, of this algorithm for specific problems, it is possible to design tailor-made dynamic programming algorithms on the corresponding tree decompositions. Treewidth has also been important from the combinatorial point of view. This is mostly due to the celebrated "*planar graph exclusion theorem*" [14,15]. This theorem asserts that:

(*) *Every graph that does not contain some fixed wall*[1] *as a topological minor*[2] *has bounded treewidth.*

The above result had a considerable algorithmic impact as every problem for which a negative (or positive) answer can be certified by the existence of some sufficiently big wall in its input, is reduced to its resolution on graphs of bounded treewidth. This induced a lot of research on the derivation of fast parameterized algorithms that can construct (optimally or approximately) these decompositions. For instance, according to [1], treewidth can be computed in $f(OPT) \cdot n$ steps where $f(w) = 2^{O(w^3)}$ while, more recently, a 5-approximation for treewidth was given in [2] that runs in $2^{O(OPT)} \cdot n$ steps.

Unfortunately, the aforementioned success stories about treewidth have some natural limitations. In fact, it is not always possible to use treewidth for improving the tractability of NP-hard problems. In particular, there are interesting cases of problems where no such an FPT-algorithm is expected to exist [6,7,10]. Therefore, it is an interesting question whether there are alternative, but still general, graph invariants that can provide tractable parameterizations for such problems.

A promising candidate in this direction is the graph invariant of *tree-cut width* that was recently introduced by Wollan in [23]. Tree-cut width can be

---

[1] We avoid the formal definition of a wall in this extended abstract. Instead, we provide the following image that, we believe, provides the necessary intuition.

[2] A graph H is a *topological minor* of a graph G if a subdivision of H is a subgraph of G.

seen as an "edge" analogue of treewidth. It is defined using a different type of decompositions, namely, tree-cut decompositions that are roughly tree-like partitions of a graph into mutually disjoint pieces such that both the size of some "essential" extension of these pieces and the number of edges crossing two neighboring pieces are bounded (see Sect. 2 for the formal definition). Our first result is that it is NP-hard to decide, given a graph G and an integer w, whether the input graph G has tree-cut width at most w. This follows from a reduction from the MIN BISECTION problem that is presented in Subsect. 2.2. This encourages us to consider a parameterized algorithm for this problem.

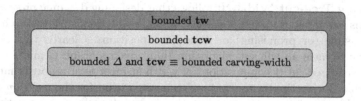

Fig. 1. The relations between classes with bounded treewidth (**tw**) and tree-cut width (**tcw**).

Another tree-like parameter that can be seen as an edge-counterpart of tree-width is *carving-width*, defined in [18]. It is known that a graph has bounded carving-width if and only if both its treewidth and its maximum degree are bounded. We stress that this is not the case for tree-cut width, which can also capture graphs with unbounded maximum degree and, thus, is more general than carving-width. There are two reasons why tree-cut width might be a good alternative for treewidth. We expose them below.

(1) Tree-cut Width as a Parameter. From now on we denote by $\mathbf{tcw}(G)$ (resp. $\mathbf{tw}(G)$) the tree-cut width (resp. treewidth) of a graph G. As it is shown in [23] $\mathbf{tcw}(G) = O(\mathbf{tw}(G) \cdot \Delta(G))$. Moreover, in [8], it was proven that $\mathbf{tw}(G) = O((\mathbf{tcw}(G))^2)$ and in Subsect. 2.3, we prove that the latter upper bound is asymptotically tight. The graph class inclusions generated by the aforementioned relations are depicted in Fig. 1. As tree-cut width is a "larger" parameter than treewidth, one may expect that some problems that are intractable when parameterized by treewidth (known to be W[1]-hard or open) become tractable when parameterized by tree-cut width. Indeed, some recent progress on the development of a dynamic programming framework for tree-cut width (see [8]) confirms that assumption. According to [8], such problems include CAPACITATED DOMINATING SET problem, CAPACITATED VERTEX COVER [6], and BALANCED VERTEX-ORDERING problem. We expect that more problems will fall into this category.

(2) Combinatorics of Tree-cut Width. In [23] Wollan proved the following counterpart of (*):

(**) *Every graph that does not contain some fixed wall as an immersion*[3] *has bounded tree-cut width.*

Notice that (*) yields (**) if we replace "topological minor" by "immersion" and "treewidth" by "tree-cut width". This implies that tree-cut width has combinatorial properties analogous to those of treewidth. It follows that every problem where a negative (or positive) answer can be certified by the existence of a wall as an immersion, can be reduced to the design of a suitable dynamic programming algorithm for this problem on graphs of bounded tree-cut width.

Computing Tree-cut Width. It follows that designing dynamic programming algorithms on tree-cut decompositions might be a promising task when this is not possible (or promising) on tree-decompositions. Clearly, this makes it imperative to have an efficient algorithm that, given a graph G and an integer w, constructs tree-cut decompositions of width at most w or reports that this is not possible. Interestingly, an $f(w) \cdot n^3$-time algorithm for the *decision version* of the problem is known to *exist* but this is not done in a constructive way. Indeed, for every fixed w, the class of graphs with tree-cut width at most w is closed under immersions [23]. By the fact that graphs are well-quasi-ordered under immersions [16], for every w, there exists a *finite* set $\mathcal{R}_w$ of graphs such that G has tree-cut width at most w if and only if it does not contain any of the graphs in $\mathcal{R}_w$ as an immersion. From [11], checking whether an h-vertex graph H is contained as an immersion in some n-vertex graph G can be done in $f(w) \cdot n^3$ steps. It follows that, for every fixed w, there *exists* a polynomial algorithm checking whether the tree-cut width of a graph is at most w. Unfortunately, the *construction* of this algorithm requires the knowledge of the set $\mathcal{R}_w$ for every w, which is not provided by the results in [16]. Even if we knew $\mathcal{R}_w$, it is not clear how to construct a tree-cut decomposition of width at most w, if one exists.

In this paper we make a first step towards a constructive parameterized algorithm for tree-cut width by giving an FPT 2-approximation for it. Given a graph G and an integer w, our algorithm either reports that G has tree-cut width more than w or outputs a tree-cut decomposition of width at most $2w$ in $2^{O(w^2 \log w)} n^2$ steps. The algorithm is presented in Sect. 3.

2 Problem Definition and Preliminary Results

Unless specified otherwise, every graph in this paper is undirected and loopless and may have multiple edges. By $V(G)$ and $E(G)$ we denote the vertex set and the edge set, respectively, of a graph G. Given a vertex $x \in V(G)$, the *neighborhood* of x is $N(x) = \{y \in V(G) \mid xy \in E(G)\}$. Given two disjoint sets X and Y of $V(G)$, we denote $\delta_G(X, Y) = \{xy \in E(G) \mid x \in X, y \in Y\}$. For a subset X of $V(G)$, we define $\partial_G(X) = \{x \in X \mid N(x) \setminus X \neq \emptyset\}$.

---

[3] A graph H is an *immersion* of a graph G if H can be obtained from some subgraph of G after replacing edge-disjoint paths with edges.

2.1   Tree-cut Width and Treewidth

Tree-cut width. A *tree-cut decomposition* of G is a pair $(T, \mathcal{X})$ where T is a tree and $\mathcal{X} = \{X_t \subseteq V(G) \mid t \in V(T)\}$ such that

- $X_t \cap X_{t'} = \emptyset$ for all distinct t and t' in $V(T)$,
- $\bigcup_{t \in V(T)} X_t = V(G)$.

From now on we refer to the vertices of T as *nodes*. The sets in $\mathcal{X}$ are called the *bags* of the tree-cut decomposition. Observe that the conditions above allow to assign an empty bag for some node of T. Such nodes are called *trivial nodes*. Observe that we can always assume that trivial nodes are internal nodes.

Let $L(T)$ be the set of leaf nodes of T. For every tree-edge $e = \{u, v\}$ of $E(T)$, we let T_u and T_v be the subtrees of $T \setminus e$ which contain u and v, respectively. We define the *adhesion* of a tree-edge $e = \{u, v\}$ of T as follows:

$$\delta^T(e) = \delta_G \Big(\bigcup_{t \in V(T_u)} X_t, \bigcup_{t \in V(T_v)} X_t \Big).$$

For a graph G and a set $X \subseteq V(G)$, the *3-center* of (G, X) is the graph obtained from G by repetitively dissolving every vertex $v \in V(G) \setminus X$ that has two neighbors and degree 2 and removing every vertex $w \in V(G) \setminus X$ that has degree at most 2 and one neighbor (*dissolving* a vertex x of degree two with exactly two neighbors y and z is the operation of removing x and adding the edge $\{y, z\}$ – if this edge already exists then its multiplicity is increased by one). Given a tree-cut decomposition $(T, \mathcal{X})$ of G and node $t \in V(T)$, let $T_1, \ldots, T_\ell$ be the connected components of $T \setminus t$. The *torso* of G *at* t, denoted by H_t, is a graph obtained from G by identifying each non-empty vertex set $Z_i := \bigcup_{b \in V(T_i)} X_b$ into a single vertex z_i (in this process, parallel edges are kept). We denote by $\bar{H}_t$ the 3-center of (H_t, X_t). Then the *width* of $(T, \mathcal{X})$ equals

$$\max \left(\{|\delta^T(e)| : e \in E(T)\} \cup \{|V(\bar{H}_t)| : t \in V(T)\} \right).$$

The *tree-cut width* of G, or $\mathbf{tcw}(G)$ in short, is the minimum width of $(T, \mathcal{X})$ over all tree-cut decompositions $(T, \mathcal{X})$ of G.

The following definitions will be used in the approximation algorithm. Let $(T, \mathcal{X})$ be a tree-cut decomposition of G. It is *non-trivial* if it contains at least two non-empty bags, and *trivial* otherwise. We will assume that every leaf of a tree-cut decomposition has a non-empty bag. The *internal-width* of a non-trivial tree-cut decomposition $(T, \mathcal{X})$ is

$$\mathbf{in\text{-}tcw}(T, \mathcal{X}) = \max \left(\{|\delta^T(e)| : e \in E(T)\} \cup \{|V(\bar{H}_t)| : t \in V(T) \setminus L(T)\} \right).$$

If $(T, \mathcal{X})$ is trivial, then we set $\mathbf{in\text{-}tcw}(T, \mathcal{X}) = 0$.

We decision problem corresponding to tree-cut width is the following:

TREE-CUT WIDTH
Input: a graph G and a non-negative integer k.
Question: $\mathbf{tcw}(G) \le k$?

Treewidth. A *tree decomposition* of a graph G is a pair $(T, \mathcal{Y}) = \{Y_x : x \in V(T)\}$) such that T is a tree and $\mathcal{Y}$ is a collection of subsets of $V(G)$ where

- $\bigcup_{x \in V(T)} Y_x = V(G)$;
- for every edge $\{u, v\} \in E(G)$ there exists $x \in V(T)$ such that $u, v \in Y_x$; and
- for every vertex $u \in V(G)$ the set of nodes $\{x \in V(T) : u \in Y_x\}$ induces a subtree of T.

The vertices of T are called *nodes* of $(T, \mathcal{Y})$ and the sets Y_x are called bags. The *width* of a tree decomposition is the size of the largest bag minus one. The *treewidth* of a graph, denoted by $\mathbf{tw}(G)$, is the smallest width of a tree decomposition of G.

2.2 Computing Tree-cut Width Is NP-complete

We prove that TREE-CUT WIDTH is NP-hard by a polynomial-time reduction from MIN BISECTION, which is known to be NP-hard [9]. The input of MIN BISECTION is a graph G and a non-negative integer k, and the question is whether there exists a bipartition (V_1, V_2) of $V(G)$ such that $|V_1| = |V_2|$ and $|\delta_G(V_1, V_2)| \leqslant k$.

Theorem 1 ($\star$). TREE-CUT WIDTH *is NP-complete.*

2.3 Tree-cut Width Vs Treewidth

In this section we investigate the relation between treewidth and tree-cut width. The following was proved in [8].

Proposition 1. *For a graph of tree-cut width at most w, its treewidth is at most $2w^2 + 3w$.*

We now prove that the bound of Proposition 1 is asymptotically optimal.

We define the family of graphs $\mathcal{H} = \{H_w : w \in \mathbb{N}_{\geqslant 1}\}$ as follows. The vertex set of H_w is a disjoint union of w cliques, $Q_1, \ldots, Q_w$, each containing w vertices. For each $1 \leqslant i \leqslant w$, the vertices of Q_i are labeled as (i, j), $1 \leqslant j \leqslant w$. Besides the edges lying inside the cliques Q_i's, we add an edge between $(i, j) \in Q_i$ and $(j, i) \in Q_j$ for every $1 \leqslant i < j \leqslant w$. Notice that the vertex (i, i) does not have a neighbor outside Q_i. The graph H_4 is depicted in Fig. 2.

Lemma 1. *The tree-cut width of H_w is at most $w + 1$.*

PROOF: Consider the tree-cut decomposition $(T, \mathcal{X})$, in which T is a star with t as the center and $q_1, \ldots, q_w$ as leaves. For the bags, we set $X_t = \emptyset$, and $X_{q_i} = Q_i$ for $1 \leqslant i \leqslant w$. It is straightforward to verify that the tree-cut width of $(T, \mathcal{X})$ is $w + 1$. $\square$

We need some definitions that will be used in the proof of the next lemma. Let G be a graph. Two subgraphs X and Y of G *touch* each other if either

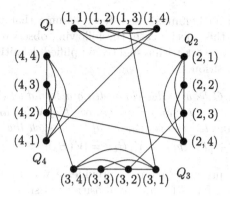

Fig. 2. The graph H_4.

$V(X) \cap V(Y) \neq \emptyset$ or there is an edge $e = \{x, y\} \in E(G)$ with $x \in V(X)$ and $y \in V(Y)$. A *bramble* $\mathcal{B}$ is a collection of connected subgraphs of G pairwise touching each other. The *order* of a bramble $\mathcal{B}$ is the minimum size of a hitting set S of $\mathcal{B}$, that is a set $S \subseteq V(G)$ such that for every $B \in \mathcal{B}$, $S \cap V(B) \neq \emptyset$. In Seymour and Thomas [17], it is known that the treewidth of a graph equals the maximum order over all brambles of G minus one. Therefore, a bramble of order k is a certificate that the treewidth is at least $k - 1$.

Lemma 2 ($\star$). *For any positive integer w, the treewidth of $H_w \in \mathcal{H}$ is at least $\frac{1}{16} w^2 - 1$.*

From Lemmas 1 and 2, we conclude to the following.

Theorem 2. *For every $w \in \mathbb{N}_{\geqslant 1}$ there exists a graph H_w such that $\mathbf{tw}(H_w) = \Omega((\mathbf{tcw}(H_w))^2)$.*

3   The 2-Approximation Algorithm

We present a 2-approximation of TREE-CUT WIDTH running in time $2^{O(w^2 \log w)} \cdot n^2$. As stated in Lemma 3 below, we first observe that computing the tree-cut width of G reduces to computing the tree-cut width of 3-edge-connected graphs. This property can be easily derived from [23, Lemmas 10–11].

Lemma 3 ($\star$). *Given a connected graph G, let $\{V_1, V_2\}$ be a partition of $V(G)$ such that $\delta_G(V_1, V_2)$ is a minimal cut of size at most two and let $w \geqslant 2$ be a positive integer. For $i = 1, 2$, let G_i be the graph obtained from G by identifying the vertex set V_{3-i} into a single vertex v_{3-i}. Then G has tree-cut width at most w if and only if both G_1 and G_2 have tree-cut width at most w.*

The proof of the next lemma is easy and is omitted.

Lemma 4. *Let G be a graph and let v be a vertex of G with degree 1 (resp. 2). Let also G' be the graph obtained from G after removing (resp. dissolving) v. Then $\mathbf{tcw}(G) = \mathbf{tcw}(G')$.*

From now on, based on Lemmas 3 and 4, we assume that the input graph is 3-edge-connected. In this special case, the following observation is not difficult to verify. It allows us to work with a slightly simplified definition of the 3-centers in a tree-cut decomposition.

Observation 1. *Let G be a 3-edge-connected graph and let $(T, \mathcal{X})$ be a tree-cut decomposition of G. Consider an arbitrary node t of $V(T)$ and let $\mathcal{T}$ be the set containing every connected component T' of $T \setminus t$ such that $\bigcup_{s \in V(T')} X_s \neq \emptyset$. Then $|V(\bar{H}_t)| = |X_t| + |\mathcal{T}|$, that is $|V(\bar{H}_t)| = |V(H_t)|$.*

We observe that the proof of Lemma 3 provides a way to construct a desired tree-cut decomposition for G from decompositions of smaller graphs. Given an input graph G for TREE-CUT WIDTH, we find a minimal cut (V_1, V_2) with $|\delta(V_1, V_2)| \leqslant 2$ and create a graph G_i as in Lemma 3, with the vertex v_{3-i} marked as distinguished. We recursively find such a minimal cut in the smaller graphs created until either one becomes 3-edge-connected or has at most w vertices.

Therefore, a key feature of an algorithm for TREE-CUT WIDTH lies in how to handle 3-edge-connected graphs. Our algorithm iteratively refines a tree-cut decomposition $(T, \mathcal{X})$ of the input graph G and either guarantees that the following invariant is satisfied or returns that $\mathbf{tcw}(G) > \omega$.

Invariant: $(T, \mathcal{X})$ *is a tree-cut decomposition of G where* $\mathbf{in\text{-}tcw}(T, \mathcal{X}) \leqslant 2 \cdot w$.

Clearly the trivial tree-cut decomposition satisfies the *Invariant*. A leaf t of T such that $|X_t| \geqslant 2 \cdot \omega$ is called a *large leaf*. At each step, the algorithm picks a large leaf and refines the current tree-cut decomposition by breaking this leaf bag into smaller pieces. The process repeats until we finally obtain a tree-cut decomposition of width at most $2w$, or encounter a certificate that $\mathbf{tcw}(G) > w$.

3.1 Refining a Large Leaf of a Tree-cut Decomposition

A large leaf will be further decomposed into a star. To that aim, we will solve the following problem:

CONSTRAINED STAR-CUT DECOMPOSITION
Input: An undirected graph G, an integer $w \in \mathbb{N}$, a set $B \subseteq V(G)$, and a weight function $\gamma : B \to \mathbb{N}$.
Parameter: w.
Output: A non-trivial tree-cut decomposition $(T, \mathcal{X})$ of G such that

1. T is a star with central node t_c and with ℓ leaves for some $\ell \in \mathbb{N}^+$,
2. $\mathbf{in\text{-}tcw}(T, \mathcal{X}) \leqslant w$, and
3. $|X_{t_c}| + \ell \leqslant w$ and for every leaf node t, $\gamma(B \cap X_t) \leqslant w$,

or report that such a tree-cut decomposition does not exist.

Observe that a YES-instance satisfies, for every $x \in B$, $\gamma(x) \leqslant w$. We also notice that as the output of the algorithm is a non-trivial tree-cut decomposition, T contains at least two nodes with non-empty bags and every leaf node is non-empty.

Given a subset $S \subseteq V(G)$, we define the instance of the CONSTRAINED STAR-CUT DECOMPOSITION problem $I(S, G) = (G[S], w, \partial_G(S), \gamma_S)$ where for every $x \in \partial_G(S)$, $\gamma_S(x) = |\delta_G(\{x\}, V(G) \setminus S)|$.

Lemma 5 ($\star$). *Let G be a 3-edge-connected graph, $w \in \mathbb{Z}_{\geqslant 2}$, and let $S \subseteq V(G)$ be a set of vertices such that $|S| \geqslant w + 1$ and $|\delta_G(S, V(G) \setminus S)| \leqslant 2w$. If $\mathrm{tcw}(G) \leqslant w$, then $I(S, G) = (G[S], w, \partial_G(S), \gamma_S)$ is a YES-instance of CONSTRAINED STAR-CUT DECOMPOSITION.*

Given a 3-edge-connected graph, applying Lemma 5 on a large leaf of a tree-cut decomposition that satisfies the *Invariant*, we obtain:

Corollary 1. *Let G be a 3-edge-connected graph G such that $\mathrm{tcw}(G) \leqslant w$, and let t be a large leaf of a tree-cut decomposition $(T, \mathcal{X})$ satisfying the Invariant. Then $I(X_t, G) = (G[X_t], w, \partial_G(X_t), \gamma_{X_t})$ is a YES-instance of CONSTRAINED STAR-CUT DECOMPOSITION.*

The next lemma shows that if a large leaf bag X_t of a tree-cut decomposition $(T, \mathcal{X})$ satisfying the *Invariant* defines a YES-instance of the **Constraint Tree-Cut Decomposition** problem, then $(T, \mathcal{X})$ can be further refined.

Lemma 6 ($\star$). *Let G be a 3-edge-connected graph G and $(T, \mathcal{X})$ be tree-cut decomposition of satisfying the Invariant. If $(T^*, \mathcal{X}^*)$ is a solution of CONSTRAINED STAR-CUT DECOMPOSITION on the instance $I(X_t, G) = (G[X_t], w, \partial_G(X_t), \gamma_{X_t})$ where t is a large leaf of $(T, \mathcal{X})$, then the pair $(\tilde{T}, \tilde{\mathcal{X}})$ where*

- $V(\tilde{T}) = (V(T) \setminus \{t\}) \cup V(T^*)$,
- $E(\tilde{T}) = (E(T) \setminus \{(t, t')\}) \cup E(T^*) \cup \{(t_c, t')\}$, *where t' is the unique neighbor of t in T and t_c is the central node of T^*,*
- $\tilde{\mathcal{X}} = (\mathcal{X} \setminus \{X_t\}) \cup \mathcal{X}^*$

is a tree-cut decomposition of G satisfying the Invariant. Moreover the number of non-empty bags is strictly larger in $(\tilde{T}, \tilde{\mathcal{X}})$ than in $(T, \mathcal{X})$.

3.2   An FPT Algorithm for CONSTRAINED STAR-CUT DECOMPOSITION

Lemma 1 provides a quadratic bound on the treewidth of a graph in term of its tree-cut width. This allows us to develop a dynamic programming algorithm for solving CONSTRAINED STAR-CUT DECOMPOSITION on graphs of bounded treewidth. To obtain a tree-decomposition, we use the 5-approximation FPT-algorithm of the following proposition.

Proposition 2 (see [2]). *There exists an algorithm which, given a graph G and an integer k, either correctly decides that $\mathrm{tw}(G) > w$ or outputs a tree-decomposition of width at most $5w + 4$ in time $2^{O(w)} \cdot n$.*

If $\mathbf{tcw}(G) \leqslant w$, then by Lemma 1 $\mathbf{tw}(G) \leqslant 2w^2 + 3w$. From Proposition 2, we may assume that G has treewidth $O(w^2)$ and, based on this and the next lemma, solve CONSTRAINED STAR-CUT DECOMPOSITION in $2^{O(w^2 \cdot \log w)} \cdot n$ steps.

A *rooted tree decomposition* $(T, \mathcal{X}, r)$ is a tree decomposition with a distinguished node r selected as the *root*. A *nice tree decomposition* $(T, \mathcal{Y}, r)$ (see [13]) is a rooted tree decomposition where T is binary, the bag at the root is $\emptyset$, and for each node x with two children y, z it holds $Y_x = Y_y = Y_z$, and for each node x with one child y it holds $Y_x = Y_y \cup \{u\}$ or $Y_x = Y_y \setminus \{u\}$ for some $u \in V(G)$. Notice that a nice tree decomposition is always a rooted tree decomposition. We need the following proposition.

Proposition 3 (see [1]). *For any constant $k \geqslant 1$, given a tree decomposition of a graph G of width $\leqslant k$ and $O(|V(G)|)$ nodes, there exists an algorithm that, in $O(|V(G)|)$ time, constructs a nice tree decomposition of G of width $\leqslant k$ and with at most $4|V(G)|$ nodes.*

Lemma 7 ($\star$). *Let (G, w, B, γ) be an input of* CONSTRAINED STAR-CUT DECOMPOSITION *and let $\mathbf{tw}(G) \leqslant q$. There exists an algorithm that given (G, w, B, γ) outputs, if one exists, a solution of (G, w, B, γ) in $2^{O((q+w) \log w)} \cdot n$ steps.*

3.3 Piecing Everything Together

We now present a 2-approximation algorithm for TREE-CUT WIDTH leading to the following result.

Theorem 3. *There exists an algorithm that, given a graph G and a $w \in \mathbb{Z}_{\geqslant 0}$, either outputs a tree-cut decomposition of G with width at most $2w$ or correctly reports that no tree-cut decomposition of G with width at most w exists in $2^{O(w^2 \cdot \log w)} \cdot n^2$ steps.*

PROOF: Recall that, by Lemmas 3 and 4, we can assume that G is 3-edge-connected. If not, we iteratively decompose G into 3-edge-connected components using the linear-time algorithm of [22]. A tree-cut decomposition of G can easily built from the tree-cut decomposition of its 3-edge-connected components using Lemma 3. As mentioned earlier, the trivial tree-cut decomposition satisfies the *Invariant*. Let $(T, \mathcal{X})$ be a tree-cut decomposition satisfying the *Invariant*. As long as the current tree-cut decomposition $(T, \mathcal{X})$ contains a large leaf ℓ, the algorithm applies the following steps repeatedly:

1. Let $X_\ell \in \mathcal{X}$ be the bag associated to a large leaf ℓ. Compute a nice tree-decomposition of $G[X_\ell]$ of width at most $O(w^2)$ in $2^{O(w^2)} \cdot n$ time. If such a decomposition does not exist, as $G[X_\ell]$ is a subgraph of G, Lemma 1 implies $\mathbf{tcw}(G) > w$ and the algorithm stops.
2. Solve CONSTRAINED STAR-CUT DECOMPOSITION on $I(X_t, G)$ using the dynamic programming of Lemma 7 for $q = O(w^2)$ in time $2^{O(w^2 \cdot \log w)} \cdot n$.

3. If $I(X_t, G)$ is a NO-instance, then by Corollary 1, $\mathbf{tcw}(G) > w$ and the algorithm stops.
4. Otherwise, by Lemma 6, $(T, \mathcal{X})$ can be refined into a new tree-cut decomposition satisfying the *Invariant*.

The algorithm either stops when we can correctly report that $\mathbf{tcw}(G) > w$ (step 1 or 3) or when the current tree-cut decomposition has no large leaf. In the latter case, as $(T, \mathcal{X})$ satisfies (*), it holds that $\mathbf{tcw}(T, \mathcal{X}) \leqslant 2 \cdot w$. Observe that each refinement step (step 4) strictly increases the number of non-empty bags (see Lemma 6). It follows that the above steps are repeated at most n times, implying that the running time of the 2-approximation algorithm is $2^{O(w^2 \cdot \log w)} \cdot n^2$. $\qquad\square$

4   Open Problems

The main open question is on the possibility of improving the running time or the approximation factor of our algorithm. Notice that the parameter dependence $2^{O(w^2 \cdot \log w)}$ is based on the fact that the tree-cut width is bounded by a quadratic function of treewidth. As we proved (Theorem 2), there is no hope of improving this upper bound. Therefore any improvement of the parametric dependence should avoid dynamic programming on tree-decompositions or significantly improve the running time. Another issue is whether we can improve the quadratic dependence on n to a linear one. In this direction we actually believe that an exact FPT-algorithm for the tree-cut width can be constructed using the "set of characteristic sequences" technique, as this was done for other width parameters [3,4,12,19–21]. However, as this technique is more involved, we believe that it would imply a higher parametric dependence than the one of our algorithm.

Acknowledgement. We would like to thank the anonymous reviewers for helpful remarks that improved the presentation of the manuscript.

References

1. Bodlaender, H.L.: A linear time algorithm for finding tree-decompositions of small treewidth. SIAM J. Comput. **25**, 1305–1317 (1996)
2. Bodlaender, H.L., Drange, P.G., Dregi, M.S., Fomin, F., Lokshtanov, D., Pilipczuk, M.: An $O(c^k n)$ 5-approximation algorithm for treewidth. In: IEEE Symposium on Foundations of Computer Science, FOCS, pp. 499–508 (2013)
3. Bodlaender, H.L., Kloks, T.: Efficient and constructive algorithms for the pathwidth and treewidth of graphs. J. Algorithms **21**(2), 358–402 (1996)
4. Bodlaender, H.L., Thilikos, D.M.: Computing small search numbers in linear time. In: Downey, R.G., Fellows, M.R., Dehne, F. (eds.) IWPEC 2004. LNCS, vol. 3162, pp. 37–48. Springer, Heidelberg (2004)
5. Courcelle, B.: The monadic second-order logic of graphs. I. Recognizable sets of finite graphs. Inf. Comput. **85**(1), 12–75 (1990)

6. Dom, M., Lokshtanov, D., Saurabh, S., Villanger, Y.: Capacitated domination and covering: a parameterized perspective. In: Grohe, M., Niedermeier, R. (eds.) IWPEC 2008. LNCS, vol. 5018, pp. 78–90. Springer, Heidelberg (2008)

7. Fellows, M.R., Fomin, F.V., Lokshtanov, D., Rosamond, F., Saurabh, S., Szeider, S., Thomassen, C.: On the complexity of some colorful problems parameterized by treewidth. Inf. Comput. **209**(2), 143–153 (2011)

8. Ganian, R., Kim, E.J., Szeider, S.: Algorithmic applications of tree-cut width. In: Italiano, G.F., Pighizzini, G., Sannella, D.T. (eds.) MFCS 2015. LNCS, vol. 9235, pp. 348–360. Springer, Heidelberg (2015)

9. Garey, M.R., Johnson, D.S.: Computers and Intractability: A Guide to the Theory of NP-Completeness. W. H. Freeman and Company, New York (1979)

10. Golovach, P.A., Thilikos, D.M.: Paths of bounded length and their cuts: Parameterized complexity and algorithms. Discrete Optim. **8**(1), 72–86 (2011)

11. Grohe, M., Kawarabayashi, K., Marx, D., Wollan, P.: Finding topological subgraphs is fixed-parameter tractable. In: ACM Symposium on Theory of Computing, STOC, pp. 479–488 (2011)

12. Jeong, J., Kim, E.J., Oum, S.: Constructive algorithm for path-width of matroids. CoRR (2015). arXiv:1507.02184

13. Kloks, T.: Treewidth: Computations and Approximations. LNCS, vol. 842. Springer, Heidelberg (1994)

14. Robertson, N., Seymour, P.D.: Graph minors. III. Planar tree-width. J. Comb. Theory Ser. B **36**(1), 49–64 (1984)

15. Robertson, N., Seymour, P.D.: Graph minors. V. Excluding a planar graph. J. Comb. Theory Ser. B **41**(1), 92–114 (1986)

16. Robertson, N., Seymour, P.D.: Graph minors. XXIII. Nash-Williams' immersion conjecture. J. Comb. Theory Ser. B **100**(2), 181–205 (2010)

17. Seymour, P.D., Thomas, R.: Graph searching and a min-max theorem for treewidth. J. Comb. Theory Ser. B **58**(1), 22–33 (1993)

18. Seymour, P.D., Thomas, R.: Call routing and the ratcatcher. Combinatorica **14**(2), 217–241 (1994)

19. Soares, R.P.: Pursuit-evasion, decompositions and convexity on graphs. PhD thesis, COATI, INRIA/I3S-CNRS/UNS Sophia Antipolis, France and ParGO Research Group, UFC Fortaleza, Brazil (2014)

20. Thilikos, D.M., Serna, M.J., Bodlaender, H.L.: Cutwidth I: A linear time fixed parameter algorithm. J. Algorithms **56**(1), 1–24 (2005)

21. Thilikos, D.M., Serna, M.J., Bodlaender, H.L.: Cutwidth II: Algorithms for partial w-trees of bounded degree. J. Algorithms **56**(1), 25–49 (2005)

22. Watanabe, T., Taoka, S., Mashima, T.: Minimum-cost augmentation to 3-edge-connect all specified vertices in a graph. In: ISCAS, pp. 2311–2314 (1993)

23. Wollan, P.: The structure of graphs not admitting a fixed immersion. J. Comb. Theory Ser. B **110**, 47–66 (2015)

Tight Bounds for Double Coverage Against Weak Adversaries

Nikhil Bansal[1], Marek Eliáš[1], Łukasz Jeż[1,2], Grigorios Koumoutsos[1]([⊠]),
and Kirk Pruhs[3]

[1] Eindhoven University of Technology, Eindhoven, The Netherlands
{n.bansal,m.elias,l.jez,g.koumoutsos}@tue.nl
[2] Institute of Computer Science, University of Wrocław, Wrocław, Poland
[3] University of Pittsburgh, Pittsburgh, USA
kirk@cs.pitt.edu

Abstract. We study the Double Coverage (DC) algorithm for the k-server problem in the (h, k)-setting, i.e. when DC with k servers is compared against an offline optimum algorithm with $h \leq k$ servers. It is well-known that DC is k-competitive for $h = k$. We prove that even if $k > h$ the competitive ratio of DC does not improve; in fact, it increases up to $h + 1$ as k grows. In particular, we show matching upper and lower bounds of $\frac{k(h+1)}{k+1}$ on the competitive ratio of DC on any tree metric.

1 Introduction

We consider the k-server problem defined as follows. There are k servers located on points of a metric space. In each step, a request arrives at some point of the metric space and must be served by moving some server to that point. The goal is to minimize the total distance travelled by the servers.

The k-server problem was defined by Manasse et al. [7] as a far reaching generalization of various online problems. The most well-studied of those is the paging (caching) problem, which corresponds to k-server on a uniform metric space. Sleator and Tarjan [8] gave several k-competitive algorithms for paging and showed that this is the best possible ratio for any deterministic algorithm.

Interestingly, the k-server problem does not seem to get harder on more general metrics and the celebrated k-server conjecture states that a k-competitive deterministic algorithm exists for every metric space. In a breakthrough result, Koutsoupias and Papadimitriou [6] showed that the work function algorithm (WFA) is $2k - 1$ competitive for every metric space, almost resolving the conjecture. The conjecture has been settled for several special metrics (an excellent reference is [2]). In particular for the line metric, Chrobak et al. [3] gave an elegant k-competitive algorithm called Double Coverage (DC). This algorithm was later extended and shown to be k-competitive for all tree metrics [4]. Additionally, in [1] it was shown that WFA is k-competitive for some special metrics, including the line.

Supported by NWO grant 639.022.211, ERC consolidator grant 617951, NCN grant DEC-2013/09/B/ST6/01538,NSF grants CCF-1115575, CNS-1253218, CCF-1421508, and an IBM Faculty Award.

© Springer International Publishing Switzerland 2015
L. Sanità and M. Skutella (Eds.): WAOA 2015, LNCS 9499, pp. 47–58, 2015.
DOI: 10.1007/978-3-319-28684-6_5

(h, k)-**Server Problem:** In this paper, we consider the (h, k)-setting, where the online algorithm has k servers, but its performance is compared to an offline optimal algorithm with $h \leq k$ servers. This is also known as the weak adversaries model [5], or the resource augmentation version of k-server. The (h, k)-server setting turns out to be much more intriguing and is much less understood.

For the uniform metric (the (h, k)-paging problem), $k/(k-h+1)$-competitive algorithms are known [8] and no deterministic algorithm can achieve a better ratio. Note that this guarantee equals k for $h = k$, and tends to 1 as the ratio of the number of online to offline servers k/h becomes arbitrarily large. The same competitive ratio can also be achieved for the weighted caching problem [9].

However, unlike for k-server, the underlying metric space seems to play a very important role in the (h, k)-setting. Bar-Noy and Schieber (see [2], page 175) showed that for the $(2, k)$-server problem on a line metric, no deterministic algorithm can be better than 2-competitive for any k. In particular, the ratio does not tend to 1 as k increases.

In fact, there is huge gap in our understanding of the problem, even for very special metrics. For example, for the line no guarantee better than h is known even when $k/h \to \infty$. On the other hand, the only lower bounds known are the result of Bar-Noy and Schieber mentioned above and a general lower bound of $k/(k-h+1)$ for any metric space with at least $k+1$ points (cf. [2] for both results). In particular, no lower bound better than 2 is known for any metric space and any $h > 2$, if we let $k/h \to \infty$. The only general upper bound is due to Koutsoupias [5], who showed that WFA is at most $2h$-competitive[1] for the (h, k)-server problem on any metric[2].

The DC Algorithm: This situation motivates us to consider the (h, k)-server problem on the line and more generally on trees. In particular, we consider the DC algorithm [3], defined as follows.

DC-Line: If the current request r lies outside the convex hull of current servers, serve it with the nearest server. Otherwise, we move the two servers adjacent to r towards it with equal speed until some server reaches r. If there are multiple adjacent servers at the same location, we move one of them arbitrarily.

DC-Tree: We move all the servers adjacent to r towards it at equal speed until some server reaches r. (Note that the set of adjacent servers can change during the move, and is constantly updated.)

There are several natural reasons to consider DC for line and trees. For paging (and weighted paging), all known k-competitive algorithms also attain the optimal ratio for the (h, k) version. This suggests that k-competitive algorithms for the k-server on the line might attain the "right" ratio for the (h, k)-setting. DC is the only (other than WFA) deterministic k-server algorithm known for

---

[1] Actually [5] shows a slightly stronger upper bound $\text{WFA}_k \leq 2h\text{OPT}_h - \text{OPT}_k + \text{const}$ where OPT_k and OPT_h are the optimal cost using k and h servers respectively.

[2] If the online algorithm knows h, it can simply disable its $k - h$ extra servers and be $2h - 1$ competitive (which is slightly better than $2h$). However, Koutsoupias (and also us) consider the setting where the online algorithm does not know h.

the line and trees. Moreover, DC obtains the optimum $k/(k-h+1)$-competitive ratio for the (h,k)-paging problem[3].

It seems plausible that WFA might perform very well for lines and trees as k increases, but no $o(h)$ bound is known. Most known upper bounds, including [5], bound the *extended cost* instead of the actual cost of the algorithm. Using this approach we can easily show that WFA is $(h+1)$-competitive for the line[4].

Our Results: We determine the exact competitive ratio of DC on lines and trees in the (h,k)-setting.

Theorem 1. *The competitive ratio of DC is at least* $\frac{k(h+1)}{(k+1)}$, *even for a line.*

Note that for a fixed h, the competitive ratio worsens as the number of online servers k increases! In particular, it equals h for $k=h$ and it approaches $h+1$ as $k \to \infty$.

Consider the (seemingly trivial) case of $h=1$. If $k=1$, clearly DC is 1-competitive. However, for $k=2$ it becomes 4/3 competitive[5]. Generalizing this example to $(1,k)$ already becomes quite involved. Our lower bound in Theorem 1 for general h and k is based on an adversarial strategy obtained by a careful recursive construction.

Next, we give a matching upper bound.

Theorem 2. *For any tree, the competitive ratio of DC is at most* $\frac{k(h+1)}{(k+1)}$.

This generalizes the previous results for $h=k$ [3,4]. Our proof also follows similar ideas, but our potential function is more involved (it has three terms instead of two) and the analysis is more subtle. To keep the main ideas clear, we first prove Theorem 2 for the simpler case of a line in Sect. 3. The proof for trees is analogous but more involved, and is described in Sect. 4.

2   Lower Bound

We now prove Theorem 1. We will describe an adversarial strategy S_k for the setting where DC has k servers and the offline optimum (adversary) has h servers and then show that the competitive ratio of DC can be made arbitrarily close to $k(h+1)/(k+1)$.

Roughly speaking (and ignoring some details), the strategy S_k works as follows. Let $I = [0, b_k]$ be the *working interval* associated with S_k. Let $L = [0, \epsilon b_k]$

---

[3] Here, we view the uniform metric as a star graph where requests appear to the leaves. A proof of this result will be given in the full version of the paper.

[4] In [1] it is shown that for the line $\text{ExtCost}_h \leq (h+1)\,\text{OPT}_h+$ const. Moreover in [5] the monotonicity of extended cost was proven: $\text{ExtCost}_k \leq \text{ExtCost}_h$. Using same arguments as in [5] it follows that $\text{WFA}_k \leq (h+1)\text{OPT}_h- \text{OPT}_k+$ const.

[5] Consider the instance where all servers are at $x=0$ initially. A request arrives at $x=2$, upon which both DC and offline move a server there and pay 2. Then a request arrives at $x=1$. DC moves both servers there and pays 2 while offline pays 1. All servers are now at $x=1$ and the instance repeats.

and $R = [(1-\epsilon)b_k, b_k]$ denote the (tiny) *left front* and *right front* of I. Initially, all offline and online servers are located in L. The adversary moves all its h servers to R and starts requesting points in R, until DC eventually moves all its servers to R. The strategy inside R is defined recursively depending on the number of DC servers currently in R. Roughly, if DC has i servers in R, the adversary executes the strategy S_i repeatedly inside R, until another DC server moves there, at which point it switches to the strategy S_{i+1}. When all DC servers reach R, the adversary moves all its h servers back to L and repeats the symmetric version of the above instance until all servers move from R to L. This defines a *phase*. To show the desired lower bound, we recursively bound the online and offline costs during a phase of S_k in terms of costs incurred by strategies $S_1, S_2, \ldots, S_{k-1}$.

A crucial parameter of a strategy will be the *pull*. Recall that DC moves some server q_L closer to R if and only if q_L is the rightmost DC server outside R and a request is placed to the left of q_R, the leftmost DC server in R, as shown in Fig. 1. In this situation q_R moves by δ to the left and q_L moves to the right by the same distance, and we say that the instance in R exerts a *pull* of δ on q_L. We will be interested in the amount of pull exerted by a strategy during one phase.

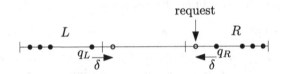

Fig. 1. DC server is pulled to the right by δ

Formal Description: We now give a formal definition of the instance. We begin by defining the following quantities associated with each strategy S_i during a single phase:

- d_i, lower bound for the cost of DC inside the working interval.
- A_i, upper bound for the cost of the adversary.
- p_i, P_i, lower resp. upper bound for the "pull" exerted on any external DC servers located to the left of the working interval of S_i. Note that, as will be clear later, by symmetry the same pull is exerted to the right.

For $i \geq h$, the ratio $r_i = \frac{d_i}{A_i}$ is a lower bound for the competitive ratio of DC with i servers against adversary with h servers.

We now define the right and left front precisely. Let $\varepsilon > 0$ be a sufficiently small constant. For $i \geq h$, we define the size of working intervals for strategy S_i as $s_h := h$ and $s_{i+1} := s_i/\varepsilon$. Note that $s_k = h/\varepsilon^{k-h}$. The working interval for strategy S_k is $[0, s_k]$ and inside it we have two working intervals for strategies S_{k-1}: $[0, s_{k-1}]$ and $[s_k - s_{k-1}, s_k]$. We continue this construction recursively and the nesting of these intervals creates a tree-like structure as shown in Fig. 2. For $i \geq h$, the working intervals for strategy S_i are called type-i intervals. Strategies S_i, for $i \leq h$, are special and are executed in type-h intervals.

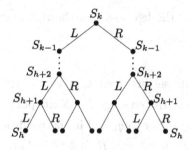

Fig. 2. Representation of strategies and the areas that they define using a binary tree.

Strategies S_i for $i \leq h$: For $i \leq h$, strategies S_i are performed in a type-h interval (recall this has length h). Let Q be $h+1$ points in such an interval, with distance 1 between consecutive points.

There are two variants of S_i that we call $\overrightarrow{S_i}$ and $\overleftarrow{S_i}$. We describe $\overrightarrow{S_i}$ in detail, and the construction of $\overleftarrow{S_i}$ will be exactly symmetric. At the beginning of $\overrightarrow{S_i}$, we will ensure that DC servers occupy the rightmost i points of Q and offline servers occupy the rightmost h points of Q as shown in Fig. 3. The adversary requests the sequence $q_{i+1}, q_i, \ldots, q_1$. It is easily verified that DC incurs cost $d_i = 2i$, and its servers will return to the initial position $q_i, \ldots, q_1$, so we can iterate $\overrightarrow{S_i}$ again. Moreover, a pull of $p_i = 1 = P_i$ is exerted in both directions.

For $i < h$, the adversary does not have to move at all, thus $A_i = 0$. For $i = h$, the offline can serve the sequence with cost $A_h = 2$, by using the server in q_h to serve request in q_{h+1} and then moving it back to q_h.

For strategy $\overleftarrow{S_i}$ we just number the points of Q in the opposite direction (q_1 will be leftmost and q_{h+1} rightmost). The request sequence, analysis, and assumptions about initial position are the same.

points of Q
servers of adversary
servers of DC

Fig. 3. Strategy $\overrightarrow{S_3}$, where $h \geq 3$.

Strategies S_i for $i > h$: We define the strategy S_i for $i > h$, assuming that $S_1, \ldots, S_{i-1}$ are defined. Let I denote the working interval for S_i. We assume that, initially, all offline and DC servers lie in the leftmost (or analogously rightmost) type-$(i-1)$ interval of I. Indeed, for S_k this is achieved by the initial configuration, and for $i < k$ we will ensure this condition before applying strategy S_i. In this case our phase consists of left-to-right step followed by right-to-left step (analogously, if all servers start in the rightmost interval, we apply first right-to-left step followed by left-to-right step to complete the phase).

Let L_j and R_j denote the leftmost and the rightmost type-j interval contained in I, for $h \leq j < i$.

Left-to-right step:

1. Adversary moves all its servers from L_{i-1} to R_h, specifically to the points $q_1, \ldots, q_h$ to prepare for the strategy $\overrightarrow{S_1}$. Next, point q_1 is requested which forces DC to move one server to q_1 and initial conditions of $\overrightarrow{S_1}$ are satisfied.

2. For $j = 1$ to h: apply $\overrightarrow{S_j}$ to interval R_h until $(j+1)$-th server arrives to point q_{j+1} in R_h. After server $j+1$ arrives, we finish the already started request sequence of $\overrightarrow{S_j}$, so that DC servers will be lined in points $q_{j+1}, \ldots, q_1$ — ready for strategy $\overrightarrow{S_{j+1}}$.

3. For $h < j < i$: apply $\overrightarrow{S_j}$ to interval R_j until $(j+1)$-th server arrives to R_j. Note that it was the only DC server moving from L_{i-1} towards R_j. The rest are either still in L_{i-1} or in R_j. Since R_j is the rightmost interval of R_{j+1} and $L_{i-1} \cap R_{j+1} = \emptyset$, our configuration is ready for strategy $\overrightarrow{S_{j+1}}$.

Right-to-left step: Same as Left-to-right, just replace $\overrightarrow{S_j}$ by $\overleftarrow{S_j}$, R_j intervals by L_j, and L_j by R_j.

Bounding Costs: We begin with a simple but useful observation that follows directly from the definition of DC. For any subset X of $i \leq k$ consecutive DC servers, let us call *center of mass* of X the average position of servers in X. We call a request *external* with respect to X, when it is outside the convex hull of X and *internal* otherwise.

Lemma 1. *For any sequence of internal requests with respect to X, the center of mass of X remains the same.*

Proof. Follows trivially since for any internal request, DC moves precisely two servers by an equal amount in opposite directions.   □

Let us derive values d_i, A_i, p_i, and P_i assuming that they were already computed for all $j < i$. We claim that the offline cost A_i for strategy S_i during a phase can be upper bounded as follows.

$$A_i \leq 2\left(s_i h + \sum_{j=1}^{i-1} A_j \frac{s_i}{p_j}\right) = 2s_i\left(h + \sum_{j=h}^{i-1} \frac{A_j}{p_j}\right) \tag{1}$$

The term $2s_i h$ follows as offline initially moves the h serves from left of I to right of I and the then back. The costs $A_j \frac{s_i}{p_j}$ are incurred during the phases S_j for $j = 1, \ldots, i-1$, because A_j is an upper bound on offline cost during a phase of strategy S_j and $\frac{s_i}{p_j}$ is an upper bound on the number of iterations of S_j during S_i. This follows because S_j (during left to right phase) executes as long as the $(j+1)$-th server moves from left of I to right of I. It travels distance at most s_i

and feels a pull of p_j while S_j is executed in R. The equality above follows, as $A_j = 0$ for $j < h$.

We now lower bound the cost of DC. Let us denote $\delta := (1 - 2\varepsilon)$. The length of $I \setminus (L_{i-1} \cup R_{i-1})$ is δs_i and all DC servers moving from right to left have to travel at least this distance. Furthermore, as $\frac{\delta s_j}{P_j}$ is a lower bound for the number of iterations of strategy S_j, we obtain:

$$d_i \geq 2\left(\delta s_i i + \sum_{j=1}^{i-1} d_j \frac{\delta s_i}{P_j}\right) = 2\delta s_i\left(i + \sum_{j=1}^{i-1} \frac{d_j}{P_j}\right) \tag{2}$$

It remains to show the upper and lower bounds on the pull P_i and p_i exerted on external servers due to the (right-to-left step of) strategy S_i. Suppose S_i is executing in interval I. Let x denote the closest DC server strictly to the left of I. Let X denote the set containing x and all DC servers located in I. The crucial point is, that during the right-to-left step of S_i all requests are internal with respect to X. So by Lemma 1, the center of the mass of X stays unchanged. As i servers moved from right to left during right-to-left step of S_i, this implies that q should have been pulled to the left by the same total amount, which is at least $i\delta s_i$ and at most is_i.

$$P_i := is_i \qquad\qquad p_i := i\delta s_i \tag{3}$$

Due to a symmetric argument, during the left-to-right step, the same amount of pull is exerted to the right.

Proof (of Theorem 1). The proof is by induction. In particular, for each $i \in [h, k]$ we will show inductively that

$$\frac{d_i}{P_i} \geq 2i\delta^{i-h} \quad \text{and} \qquad \frac{A_i}{p_i} \leq \frac{2(i+1)}{h+1}\delta^{-(i-h)} \tag{4}$$

Setting $i = k$, this implies the theorem as the competitive ratio r_k satisfies

$$r_k \geq \frac{d_k}{A_k} \geq \frac{d_k/P_k}{A_k/p_k} \geq \frac{2k}{\frac{2(k+1)}{h+1}} \frac{\delta^{k-h}}{\delta^{-(k-h)}} = \frac{k(h+1)}{k+1}\delta^{2(k-h)}$$

Choosing $\varepsilon \ll 1/(k-h)$ small enough, $\delta = (1 - 2\varepsilon)$ can be made arbitrarily close to 1, which implies the result.

Induction base $i = h$. For the base case we have the exact values of a_h and d_h, and, in particular, $\frac{d_h}{P_h} = 2h$ and $\frac{A_h}{p_h} = 2$.

Induction step $i > h$. Using (1), (2), and (3) we obtain:

$$\frac{d_i}{P_i} = \frac{2\delta}{i}\left(i + \sum_{j=1}^{i-1} \frac{d_j}{P_j}\right) \geq \frac{2\delta}{i}\left(i + \sum_{j=1}^{i-1} 2j\delta^{j-h}\right) \geq \frac{2\delta}{i}\delta^{i-1-h}(i + i(i-1)) = 2i\delta^{i-h}$$

$$\frac{A_i}{p_i} = \frac{2}{i\delta}\left(h + \sum_{j=h}^{i-1}\frac{A_j}{p_j}\right) \leq \frac{2}{i\delta}\left(h + \sum_{j=h}^{i-1}\frac{2(j+1)}{h+1}\delta^{-(j-h)}\right)$$

$$\leq \frac{2}{i\delta}\delta^{-(i-1-h)}\left(\frac{h(h+1) + 2\sum_{j=h}^{i-1}(j+1)}{h+1}\right)$$

$$= \frac{2}{i\delta^{i-h}}\frac{i(i+1)}{h+1} = \frac{2(i+1)}{h+1}\delta^{-(i-h)}$$

The last inequality follows as $2\sum_{j=h}^{i-1}(j+1) = i(i+1) - h(h+1)$. □

3   Upper Bound

In this section, we give an algorithm that matches the lower bound from the previous section. By OPT we denote the optimal offline algorithm.

Let r be a request issued at time t. Let X denote the configuration of DC (i.e. the set of points in the line where DC servers are located) and Y the configuration of OPT before serving request r. Similarly, let X' and Y' be the corresponding configurations after serving r. In order to prove our upper bound, we define a potential function $\Phi(X,Y)$ such that

$$DC(t) + \Phi(X',Y') - \Phi(X,Y) \leq c \cdot OPT(t), \tag{5}$$

where $c = \frac{k(h+1)}{k+1}$ is the desired competitive ratio, and $DC(t)$ and $OPT(t)$ denote the cost incurred by DC and OPT at time t.

Let $M \subseteq X$ be some fixed set of h servers of DC and $\mathcal{M}(M,Y)$ denote the cost of the minimum weight perfect matching between M and Y. We denote

$$\Psi_M(X,Y) := \frac{k(h+1)}{k+1}\cdot\mathcal{M}(M,Y) + \frac{k}{k+1}\cdot D_M.$$

Here, for a set of points A, D_A denotes the sum of all $\binom{|A|}{2}$ pairwise distances between points in A. The potential function is defined as follows:

$$\Phi(X,Y) = \min_M \Psi_M(X,Y) + \frac{1}{k+1}\cdot D_X$$

$$= \min_M\left(\frac{k(h+1)}{k+1}\cdot\mathcal{M}(M,Y) + \frac{k}{k+1}\cdot D_M\right) + \frac{1}{k+1}\cdot D_X.$$

Note this generalizes the potential considered in [3] for the case of $h = k$. In that setting, all the online servers are matched and hence $D_M = D_X$ and is independent of M, and thus the potential above becomes k times that minimum cost matching between X and Y plus D_x. On the other hand in our setting, we need to select the right set M of DC servers to be matched to the offline servers based on minimizing $\Psi_M(X,Y)$.

Let us first give a useful property concerning minimizers of Ψ, which will be crucial later in our analysis. Note that $\Psi_M(X,Y)$ is not simply the best matching between X and Y, but also includes the term D_M which makes the argument slightly subtle. We prove this lemma directly for trees, since it will be also useful in the following section.

Lemma 2. *Let X and Y be the configurations of DC and OPT and consider some fixed offline server at location $y \in Y$. There exists a minimizer M of Ψ that contains some DC server x which is adjacent to y. Moreover, there is a minimum cost matching $\mathcal{M}$ between M and Y that matches x to y[6].*

Proof. Let M' be some minimizer of $\Psi_M(X,Y)$ and $\mathcal{M}'$ be some associated minimum cost matching between M' and Y. Let x' denote the online server currently matched to y in $\mathcal{M}'$ and suppose that x' is not adjacent to y. We denote x the adjacent server to y, in the path from y to x'.

We will show that we can always modify the matching (and M') without increasing the cost of Φ, so that y is matched to x. We consider two cases depending on whether x is matched or unmatched.

1. If $x \in M'$: Let us call y' the offline server which is matched to x in $\mathcal{M}'$. We swap the edges and match x to y and x' to y'. The cost of the edge connecting y in the matching reduces by exactly $d(x',y) - d(x,y) = d(x',x)$. On the other hand, the cost of the matching edge for y' increases by $d(x',y') - d(x,y') \le d(x,x')$. Thus, the new matching has no larger cost. Moreover, the set of matched servers $M = M'$ and hence $D_M = D_{M'}$, which implies that $\Psi_M(X,Y) \le \Psi_{M'}(X,Y)$.

2. If $x \notin M'$: In this case, we set $M = M' \setminus \{x'\} \cup \{x\}$ and we form $\mathcal{M}$, where y is matched to x and all other offline servers are matched to the same server as in $\mathcal{M}'$. Now, the cost of the matching reduces by $d(x',y) - d(x,y) = d(x,x')$ and $D_M \le D_{M'} + (h-1) \cdot d(x,x')$ (as the distance of each server in $M' \setminus \{x'\}$ to x can be greater than the distance to x' by at most $d(x,x')$). This gives

$$\Psi_M(X,Y) - \Psi_{M'}(X,Y) \le -\frac{(h+1)k}{k+1} \cdot d(x,x') + \frac{k(h-1)}{k+1} \cdot d(x,x')$$
$$= -\frac{2k}{k+1} \cdot d(x,x') < 0,$$

and hence $\Psi_M(X,Y)$ is strictly smaller than $\Psi_{M'}(X,Y)$. □

We are now ready to prove Theorem 2 for the line.

Proof. Recall, that we are at time t and request r is arriving. We divide the analysis into two steps: (i) the offline serves r and then (ii) the online serves it. As a consequence, whenever a server of DC serves r, we can assume that a server of OPT is already there.

For all following steps considered, M will be the minimizer of $\Psi_M(X,Y)$ in the beginning of the step. It might happen that, after change of X, Y during the

---

[6] We remark that this property does not hold (simultaneously) for every offline server, but only for a single fixed offline server y.

step, better minimizer can be found. However, upper bound for $\Delta\Psi_M(X,Y)$ is sufficient to bound the change in the first term of the potential function.

Offline moves: If offline moves one of its servers by distance d to serve r the value of $\Psi_M(X,Y)$ increases by at most $\frac{k(h+1)}{k+1}d$. As $OPT(t)=d$ and X does not change, it follows that

$$\Delta\Phi(X,Y)\le\frac{k(h+1)}{k+1}\cdot OPT(t),$$

and hence (5) holds. We now consider the second step when DC moves.

DC moves: We consider two cases depending on whether DC moves a single server or two servers.

1. Suppose DC moves its rightmost server (the leftmost server case is identical) by distance d. Let y denote the offline server at r. By Lemma 2 we can assume that y is matched to the rightmost server of DC. Thus, the cost of the minimum cost matching between M and Y decreases by d. Moreover D_M increases by exactly $(h-1)d$ (as the distance to rightmost server increases by d for all servers of DC). Thus, $\Psi_M(X,Y)$ changes by

$$-\frac{k(h+1)}{k+1}\cdot d+\frac{k(h-1)}{k+1}\cdot d=-\frac{2k}{k+1}\cdot d.$$

Similarly, D_X increases by exactly $(k-1)d$. This gives us that

$$\Delta\Phi(X,Y)\le-\frac{2k}{k+1}\cdot d+\frac{k-1}{k+1}\cdot d=-d.$$

As $DC(t)=d$, this implies that (5) holds.
2. We now consider the case when DC moves 2 servers x and x', each by distance d. Let y denote the offline server at the request r. By Lemma 2 applied to y, we can assume that M contains at least one of x or x', and that y is matched to one of them (say x) in some minimum cost matching $\mathcal{M}$ of M to Y.
   We note that D_X decreases by precisely $2d$. In particular, the distance between x and x' decreases by $2d$, and for any other server of $X\setminus\{x,x'\}$ its total distance to other servers does not change. Moreover, $DC(t)=2d$. Hence, to prove (5), it suffices to show

$$\Delta\Psi_M(X,Y)\le-\frac{k}{k+1}\cdot 2d.\tag{6}$$

To this end, we consider two sub-cases.
   (a) *Both x and x' are matched:* In this case, the cost of the matching $\mathcal{M}$ does not go up as the cost of the matching edge (x,y) decreases by d and the move of x' can increase the cost of matching by at most d. Moreover, D_M decreases by precisely $2d$ (due to x and x' moving closer). Thus, $\Delta\Psi_M(X,Y)\le-\frac{k}{k+1}\cdot 2d$, and hence (6) holds.

(b) *Only x is matched (to y) and x' is unmatched:* In this case, the cost of the matching $\mathcal{M}$ decreases by d. Moreover, D_M can increase by at most $(h-1)d$, as x can move away from each server in $M \setminus \{x\}$ by distance at most d. So

$$\Delta \Psi_M(X, Y) \leq -\frac{(h+1)k}{k+1} \cdot d + \frac{k(h-1)}{k+1} \cdot d = -\frac{2k}{k+1} \cdot d,$$

i.e., (6) holds.   □

4   Extension to Trees

We now consider tree metrics. Specifically, we prove Theorem 2. Part of the analysis carries over from the previous section. We use the same potential function as for the line. Observe that Lemma 2 holds for trees: We only used the triangle inequality and the fact that there exists a unique path between any two points.

Proof (of Theorem 2). The analysis of the step when offline moves is exactly the same as for the line. In particular, if the offline algorithm moves by distance d, only the matching cost is affected in the potential function and it can increase by at most $d \cdot k(h+1)/(k+1)$.

It remains to analyze the change in the potential caused by the moves of DC. In that case, we break down the DC move into *elementary moves*. Let us call *active* the servers adjacent to the requested point r, i.e., the ones which are moving. An elementary move ends when any server reaches either the request r or a vertex of the tree. In the latter case, another elementary move immediately follows, perhaps with a different set of active servers. We are going to prove that (5) holds for every elementary move. By summation, this implies that it holds for the entire DC move.

Consider an elementary move where q servers are moving by distance d. We need to establish some notation first: Let M be a minimizer of $\Psi_M(X, Y)$ at the beginning of the step and A be the set of active servers. Let us imagine for now, that the requested point r is the root of the whole tree. For $a \in A$ let Q_a denote the set of DC servers in the subtree below a (but including a). We set $q_a := |Q_a|$ and $h_a := |Q_a \cap M|$. Finally, let $A_M := A \cap M$.

By Lemma 2, we can assume that one of the active servers is matched to offline server in r. We get that $\mathcal{M}(M, Y)$ increases by at most $(|A_M| - 2) \cdot d$.

In order to calculate the change in D_X and D_M, it is convenient to consider the moves of active servers sequentially rather than simultaneously.

For D_X, it is clear that each $a \in A$, moves further away from $q_a - 1$ DC servers by distance d and gets closer to $k - q_a$ by the same distance. Thus, the change of D_X associated with a is $(q_a - 1 - (k - q_a))d = (2q_a - k - 1)d$. Overall,

$$\Delta D_X = \sum_{a \in A} (2q_a - k - 1)d = (2k - q(k+1)) \, d, \quad \text{as} \sum_{a \in A} q_a = k.$$

Similarly, for D_M, we first note that it can change only due to moves of servers in A_M. Specifically, each $a \in A_M$, moves further away from $h_a - 1$ matched DC servers and gets closer to the rest $h - h_a$ of them. Thus, the change of D_M associated with a is $(2h_a - h - 1)d$, so overall we have

$$\Delta D_M = \sum_{a \in A_M} (2h_a - h - 1)d \le (2h - |A_M|(h+1))\, d,$$

as $\sum_{a \in A_M} h_a \le \sum_{a \in A} h_a = h$.

Using above inequalities, we see that the change of potential is at most

$$\Delta \Phi(X, Y) \le \frac{d}{k+1} \left(k(h+1)(|A_M| - 2) + k\,(2h - |A_M|(h+1)) + (2k - q(k+1)) \right)$$

$$= \frac{d}{k+1} \left(-2k(h+1) + 2kh + (2k - q(k+1)) \right)$$

$$= \frac{d}{k+1} \left(-q(k+1) \right) = -q \cdot d\,,$$

As the cost of DC is $q \cdot d$, we get that (5) holds, which completes the proof. $\square$

References

1. Bartal, Y., Koutsoupias, E.: On the competitive ratio of the work function algorithm for the k-server problem. Theor. Comput. Sci. **324**(2–3), 337–345 (2004)
2. Borodin, A., El-Yaniv, R.: Online Computation and Competitive Analysis. Cambridge University Press, Cambridge (1998)
3. Chrobak, M., Karloff, H.J., Payne, T.H., Vishwanathan, S.: New results on server problems. SIAM J. Discrete Math. **4**(2), 172–181 (1991)
4. Chrobak, M., Larmore, L.L.: An optimal on-line algorithm for k-servers on trees. SIAM J. Comput. **20**(1), 144–148 (1991)
5. Koutsoupias, E.: Weak adversaries for the k-server problem. In: Proceedings of the 40th Symposium on Foundations of Computer Science (FOCS), pp. 444–449 (1999)
6. Koutsoupias, E., Papadimitriou, C.H.: On the k-server conjecture. J. ACM **42**(5), 971–983 (1995)
7. Manasse, M.S., McGeoch, L.A., Sleator, D.D.: Competitive algorithms for server problems. J. ACM **11**(2), 208–230 (1990)
8. Sleator, D.D., Tarjan, R.E.: Amortized efficiency of list update and paging rules. Commun. ACM **28**(2), 202–208 (1985)
9. Young, N.E.: The k-server dual and loose competitiveness for paging. Algorithmica **11**(6), 525–541 (1994)

Shortest Augmenting Paths for Online Matchings on Trees

Bartłomiej Bosek[1]([✉]), Dariusz Leniowski[2], Piotr Sankowski[2], and Anna Zych[2]

[1] Theoretical Computer Science Department,
Faculty of Mathematics and Computer Science,
Jagiellonian University, Kraków, Poland
bosek@tcs.uj.edu.pl
[2] University of Warsaw,
Warsaw, Poland

Abstract. The shortest augmenting path (SAP) algorithm is one of the most classical approaches to the maximum matching and maximum flow problems, e.g., using it Edmonds and Karp in 1972 have shown the first strongly polynomial time algorithm for the maximum flow problem. Quite astonishingly, although is has been studied for many years already, this approach is far from being fully understood. This is exemplified by the online bipartite matching problem. In this problem a bipartite graph $G = (W \uplus B, E)$ is being revealed online, i.e., in each round one vertex from B with its incident edges arrives. After arrival of this vertex we augment the current matching by using shortest augmenting path. It was conjectured by Chaudhuri et al. (INFOCOM'09) that the total length of all augmenting paths found by SAP is $O(n \log n)$. However, no better bound than $O(n^2)$ is known even for trees. In this paper we prove an $O(n \log^2 n)$ upper bound for the total length of augmenting paths for trees.

1 Introduction

The shortest augmenting path (SAP) algorithm is one of the most classical approaches to the maximum matching and maximum flow problems. Using this idea Edmonds and Karp in 1972 have shown the first strongly polynomial time algorithm for the maximum flow problem [5]. Quite astonishingly, although this idea is one of the most basic algorithmic techniques, it is far from being fully understood. It is easier to talk about it by introducing the online bipartite matching problem. In this problem a bipartite graph $G = (W \uplus B, E)$ is being revealed online, i.e., in each round one vertex from B with its incident edges arrives. After arrival of this vertex we augment this matching by using shortest augmenting path. It was conjectured by Chaudhuri et al. [4] that the total length of augmenting paths found by SAP is $O(n \log n)$. However, no better bound than $O(n^2)$ is known even for trees. Proving this conjecture would have quite striking

---

This work was supported by NCN Grant 2013/11/D/ST6/03100, ERC StG project PAAl 259515 and FET IP project MULTIPLEX 317532.

L. Sanità and M. Skutella (Eds.): WAOA 2015, LNCS 9499, pp. 59–71, 2015.
DOI: 10.1007/978-3-319-28684-6_6

consequences even for maximum flow problem, as it would show that the total length of augmenting paths in unit capacity networks in Edmonds-Karp algorithm is $O(m \log n)$. This consequence is obtained via the bipartite line graph construction that is used to reduce the max-flow problem to maximum matching problem [10]. The obtained bipartite line graph has $2m$ vertices.

Our paper contributes to the study of SAP algorithm by showing that in the case of trees the total length of all augmenting paths is bounded by $O(n \log^2 n)$. This result is obtained via the application of the heavy-light decomposition of trees [15] combined with charging technique that carefully assigns shortest augmenting paths to the structure of the tree. Although, this result seems to be restricted only to trees we be believe that it constitutes the first nontrivial progress towards resolving the above conjecture. Moreover, we actually conjecture here that trees are the worst-case examples for this problem. It seems that adding more edges can only help the SAP algorithm. In addition to that we explain why SAP is harder to analyze than other augmenting path algorithms, even though it seems way more natural.

2   Related Work

The online bipartite matching problem with augmentations has recently received increasing research attention [3,4,6,7]. There are several reasons to study this problem. First of all, it provides a simple solution to the online bipartite matching algorithms used in many modern applications such as online advertising (e.g. Google Ads) [11] or client-server assignment [4]. Secondly, they could give rise to new effective offline bipartite matching algorithms as in [3]. This new algorithm provides new insights to the old problem that was studied for decades.

In this paper we concentrate on bounding the total length of augmenting paths and not on the running time. With this respect, it was shown that if the vertices of B appear in a random order, the expected total paths' length for SAP is $O(n \log n)$ [4]. The worst-case total length of paths remains an open question even for trees. In the class of trees the authors of [4] proposed a different augmenting path algorithm that achieves total paths' length of $O(n \log n)$. On the other hand, for general bipartite graphs greedy ranking algorithm [3] guarantees $O(n\sqrt{n})$ total length of paths.

First of all, the above study of online bipartite matching with augmentations should be related to the work of Gupta et al. [7] which shows an $O(n)$ bound on the total length of paths, but allows to exceed the capacity of each server by a constant factor.

Another point of view is given by the dynamic matching algorithms. Most papers in this area consider edge updates in a general fully-dynamic model which allows for both insertions and deletions intermixed with each other. We note, however, that the exact results in this model [9,14] do not imply any bound on the number of changes to the matching. Much faster update times can be achieved by constant approximate algorithms, for example [1,13], which achieve polylogarithmic and logarithmic update times. Yet, the 2-approximation can be obtained in our setting by trivial greedy algorithm that preforms no changes at all.

Better approximation factor of $\frac{3}{2}$ was achieved by [12] in $O(\sqrt{m})$ update time, and then improved by Gupta and Peng to $(1+\varepsilon)$ in $O(\sqrt{m}\varepsilon^{-2})$ [8]. The $O(\sqrt{m})$ barrier was broken by Bernstein and Stein who gave a $(\frac{3}{2}+\varepsilon)$-approximation algorithm that achieves $O(m^{1/4}\varepsilon^{-2.5})$ update time [2]. The same paper proposes an $(1+\varepsilon)$-approximation algorithm in very fast $O(\alpha(\alpha+\log n)+\varepsilon^{-4}(\alpha+\log n)+\varepsilon^{-6})$ update time for the special case of bipartite graphs with constant arboricity. However, when allowing approximation in our model a much better results are possible. An $(1+\varepsilon)$ approximation in $O(m\varepsilon^{-1})$ total time and with $O(n\varepsilon^{-1})$ total length of paths was shown in [3].

3   Preliminaries

Let us define the matching problem we consider more formally. Let W and B be two sets of vertices over which the bipartite graph will be formed. The set W (called white vertices) is given up front to the algorithm, whereas the vertices in B (black vertices) arrive online. We denote by $G_t = \langle W \uplus B_t, E_t \rangle$ the bipartite graph after the t'th black vertex has arrived. The graph G_t is constructed online in the following manner. We start with $G_0 = \langle W \uplus B_0, E_0 \rangle = \langle W \uplus \emptyset, \emptyset \rangle$. In turn t a new vertex $b_t \in B$ together with all its incident edges $E(b_t)$ is revealed and G_t is defined as:

$$\begin{cases} E_t = E_{t-1} \cup E(b_t), \\ B_t = B_{t-1} \cup \{b_t\}; \end{cases}$$

The goal of our algorithm is to compute for each G_t the maximum size matching M_t. For simplicity we assume that we add in total $|W|$ black vertices. The final graph $G_{|W|}$ which is obtained in this process will be denoted by $G = (W \uplus B, E)$. We denote $n = |W| = |B|$ and $m = |E|$.

For every $t \in [n]$, we add orientation to edges of the graph G_t. This orientation is induced by matching M_t: the matched edges are oriented towards black vertices, while the unmatched edges are oriented towards white vertices. When a new vertex b_t arrives, we get an intermediate orientation $G_t^{\mathrm{int}} = (E_t^{\mathrm{int}}, B_t)$, where the edges of b_t are oriented towards its neighbors, and the rest of the edges is oriented according to M_{t-1}. Note that G_t^{int} and G_{t-1} differ only by one vertex b_t. Any simple directed path in G_t^{int} from b_t to some unmatched white vertex is an augmenting path. In turn t, if b_t can be matched, the edges of G_t^{int} are reoriented along augmenting path π_t chosen by the algorithm, and the resulting orientation is G_t. The unmatched white vertices are called *seeds*. We denote the set of seeds after turn t as

$$S_t = \{w \in W : wb \notin M_t \text{ for any } b \in B\}.$$

So in turn t the augmenting paths in G_t^{int} are the directed paths from b_t to some $s \in S_{t-1}$. We refer to the seed of the path π_t from turn t as s_t, where $s_t \in S_{t-1}$. We represent a path as a graph consisting of path vertices and path edges. We use the notation $v \xrightarrow{\pi} v'$ to denote that a (directed) path π starts in v and ends in v', and $v \to v'$ to denote a connection via a directed edge. We use the notation

$v \in \pi$ and $\rho \subseteq \pi$ to state that a vertex $v \in V(\pi)$ and that a path ρ is a subgraph of π, respectively. We also denote the length of a path π as $|\pi|$. Throughout the paper, when we write "at time t", what we formally mean is "in G_t^{int}".

The next thing we define is a set of vertices $\mathcal{D}_t$ called dead at time t. The set $\mathcal{D}_t$ is defined as the set of vertices in G_t^{int} that cannot reach S_{t-1} via a directed path in G_t^{int}. Observe, that if at some point there is no directed path from a vertex to a seed, never again there will be such a path. If a vertex is dead, all vertices reachable from it are dead as well. Hence, no alternating path can enter such a dead region and reorient its edges to make some vertices alive. In other words, $\mathcal{D}_t \subseteq \mathcal{D}_{t+1}$ for every time moment t. The vertices of $\mathcal{D}_t$ are called dead, while the remaining vertices are called alive.

We now define the effective degree of a black vertex b in turn t as the number of it's non-dead out-neighbors:

$$\text{degeff}_t(b) = |\Gamma_t(b) \setminus \mathcal{D}_t|$$

where $\Gamma_t(b)$ is the set of vertices v such that $b \to v$ in G_t^{int}, referred to sometimes as out-neighbours of b. In particular $\text{degeff}_t(b_t)$ is the number of all non-dead neighbors (in the undirected sense) of b_t, as all the edges adjacent to b_t are directed towards its neighbors.

Since we consider in this paper the special case when G_t is a tree at any time t, we will refer to G as T, and to G_t as T_t from now on.

4   Run-away-from-the-root Algorithm

In this section we present the algorithm for trees given in [4], whose total augmenting paths' length amounts to $O(n \log n)$. We refer to this algorithm as RAFR or as the run-away algorithm. We briefly explain why their analysis does not apply to the SAP algorithm.

The RAFR algorithm maintains a forest $\mathcal{F}$, which is exactly the set of trees composed of the edges and vertices already revealed. Each tree of the forest is rooted. Initially, the trees are all singleton vertices of W. In turn t, vertex b_t connects to some trees of $\mathcal{F}$. Three cases are distinguished:

1. $\text{degeff}_t(b_t) > 1$: in this case b_t connects at least two trees, in which there are two disjoint directed paths connecting b_t with a seed. We pick the smallest such tree and route the alternating path over there. The root of the newly connected tree is the root of the largest such tree.
2. $\text{degeff}_t(b_t) = 1$: we pick a path that minimizes the number of edges traversed towards the root. In other words, we choose a path that runs away from the root as soon as possible.
3. $\text{degeff}_t(b_t) = 0$: no path is possible, b_t immediately becomes a member of a dead region.

The analysis of the above algorithm in terms of the total length of the augmenting paths is as follows. We count, for every edge, how many times this edge is traversed via augmenting paths. The edge can be either traversed when Case 1 applies (we refer to such traversal as connecting) or when Case 2 applies (we refer to such traversal as non-connecting). The edge can be connecting-traversed no more than $\log n$ times, as each augmenting path applied in Case 1 implies that the tree containing this edge doubles its size. We now observe, that between two connecting traversals or after/before the last/first connecting traversal of an edge there can be at most two non-connecting traversals. The edge, before it changes its root (Case 1 applies again), can be traversed towards the root (and reversed the opposite direction) only once. This is because after such reversal the endpoint of the traversed edge becomes dead. As a consequence, every edge is reversed $O(\log n)$ number of times and the total length $O(n \log n)$ of all applied augmenting paths follows.

This algorithm cleverly plans the uniform distribution of work between the edges. By running away from the root, it distributes the work to the edges furthest from the root, and does not do unnecessarily pass through the edges that are closer to the root. In the following sections we analyze a shortest augmenting path algorithm, which is not as clever. In particular, there are examples where a single edge can be traversed $\Omega(\sqrt{n})$ times. Hence, the simple charging techniques for RAFR do not apply to SAP.

5   Shortest Paths on Trees

In this section we study the shortest augmenting path (SAP) algorithm, which in each turn chooses the shortest among all available augmenting paths. We start by giving an easy argument, that the total length of augmenting paths for SAP is $O(n \log n)$ if all vertices b_t satisfy $\mathrm{degeff}_t(b_t) > 1$. This shows that the difficult case is to deal with vertices of effective degree 1.

Lemma 1. *If for each $t \in [n]$ it holds that* $\mathrm{degeff}_t(b_t) > 1$, *then the total length of all augmenting paths applied by* SAP *is* $O(n \log n)$.

Proof. Due to the definition of effective degree, every vertex b_t connects at least two trees T_1 and T_2 that contain a directed path connecting b_t with a seed. Let T_1 be a smaller of the two trees. The length of the shortest path π_t from b_t to a seed is at most the size of T_1. We charge the cost of π_t to $|\pi_t|$ arbitrary vertices of T_1. During the course of the SAP algorithm, every vertex can be charged at most $\log n$ times, as each time it is charged, the size of its tree doubles. The total charge is hence $O(n \log n)$. □

The main result of this paper and the subject of the remainder of this section is the bound for the general case, stated in the following theorem.

Theorem 1. *The total length of augmenting paths applied by* SAP *is* $O(n \log^2 n)$.

In order to prove Theorem 1 we introduce a few definitions and observations. The core of our proof is the concept of a *dispatching vertex*.

Definition 1. *A black vertex b is called dispatching at time t if* $\mathrm{degeff}_t(b) > 1$ *and b is the first from b_t such vertex on π_t. In such case we write* $b = \mathrm{dis}(\pi_t)$. *If there is no such black vertex on π_t, we define s_t to be the dispatching vertex at time t. We also define, for every dispatching black vertex b, the time moment* $\mathrm{tlast}(b)$, *when b is dispatching for the last time.*

So every path π_t applied by SAP is assigned a uniquely defined dispatching vertex $\mathrm{dis}(\pi_t)$. The first observation we make is that we only have to care about suffixes of π_t's starting with $\mathrm{dis}(\pi_t)$.

Definition 2. *We define the split of* $\pi_t = \mu_t \rho_t$, *where ρ_t is the suffix of π_t such that* $\mathrm{dis}(\pi_t) \xrightarrow{\rho_t} s_t$. *Path $\mu_t = \pi_t \setminus \rho_t$ is the remaining part of π_t (a possibly empty prefix that ends in a vertex preceding* $\mathrm{dis}(\pi_t)$*). We sometimes refer to the above defined suffixes as dispatching paths.*

Lemma 2. *The total length of paths μ_t is linear in the size of the tree T, i.e.,*

$$\sum_{t \in [n]} |\mu_t| \in O(n)$$

Proof. The lemma holds due to Observation 2, proven below, which states that vertices of μ_t die at the time t when π_t is applied. With this observation it is clear that the time μ_t passes through a vertex is the last time SAP visits that vertex. So every vertex in the tree is visited by μ_t for any t at most once.   □

Observation 2. *Vertices of μ_t die at the time t when π_t is applied.*

Proof. At the time when π_t is applied, all vertices on μ_t have effective degree equal to 1, i.e., they have only one alive directed out-neighbour — their successor on μ_t. If we reverse the edges, the only chance for the vertices of μ_t to be alive is the last vertex b_t. This vertex however becomes dead, because its only alive out-neighbour is removed. As a consequence the whole path dies.   □

To bound the total length of augmenting paths π_t, it remains to bound the total length of dispatching paths: $\sum_{t \in [n]} |\rho_t|$. Observe, that $\rho_t = s_t$ if $s_t = \mathrm{dis}(\pi_t)$, so there is no need to worry about such paths. It is enough to consider the sum over all dispatching paths ρ_t that start in a black dispatching vertex, so from now on we focus our attention on those. As a consequence, our goal is to bound the following sum.

Lemma 3. *The total length of non-trivial dispatching paths is $O(n \log^2 n)$, i.e.,*

$$\sum_{\substack{t \in [n]: \\ \mathrm{dis}(\pi_t) \in B}} |\rho_t| \in O(n \log^2 n)$$

The proof of Lemma 3 constitutes of two steps presented by the following two lemmas. We first bound the total length of dispatching paths which start with a dispatching vertex $b \in B$ at a time before b is dispatching for the last time (such paths are called non-final):

Lemma 4. *The total length of non-final dispatching paths is $O(n \log n)$, i.e.,*

$$\sum_{\substack{t \in [n]: \\ b = \text{dis}(\pi_t) \in B \\ t < \text{tlast}(b)}} |\rho_t| \in O(n \log n)$$

Then we move on to bounding the sum of dispatching paths starting in a dispatching vertex $b \in B$ at a time when b is dispatching for the last time (such paths are called final):

Lemma 5. *The total length of final dispatching paths is $O(n \log^2 n)$, i.e.,*

$$\sum_{\substack{t \in [n]: \\ b = \text{dis}(\pi_t) \in B \\ t = \text{tlast}(b)}} |\rho_t| \in O(n \log^2 n)$$

The distinction between final and non-final dispatching paths is made for the sake of clarity of our proofs. We now continue with the proof of Lemma 4, stated again below:

Lemma 4 The total length of non-final dispatching paths is $O(n \log n)$, i.e.,

$$\sum_{\substack{t \in [n]: \\ b = \text{dis}(\pi_t) \in B \\ t < \text{tlast}(b)}} |\rho_t| \in O(n \log n)$$

Proof. We first observe that every time some vertex $b \in B$ is dispatching not for the first time, one of its neighbours dies. To be more specific, if $b = \text{dis}(\pi_t)$ and π_t does not start in b (what happens every but the first time b is dispatching), then $w \to b \subseteq \mu_t$ for some neighbour w of b. Based on Observation 2, the vertex w dies.

Hence, if b is a dispatching vertex for the k-th out of l times at some time moment, then it has at least $l - k + 2$ alive white out-neighbours at that time. We say that a subtree hangs in the neighbour w of b, if it is obtained by the removal of b from T and it contains w. Suppose that we discard two neighbors of b with the heaviest trees hanging in them, i.e., two heaviest neighbours. Then for $k = l - 1$ we have at least one alive neighbor, for $k = l - 2$ we have at least two alive neighbors, that is, at least one alive neighbor other than the neighbor used at $k = l - 1$, and so on. In other words, for any $k < l$ we can find a distinct, not already assigned, alive neighbor w different than the two heaviest neighbors of b. However, the size of the subtree hanging in that neighbour bounds the length of the shortest augmenting path starting at b. Therefore, we can bound the total length of non-final paths dispatching at b by the total size of all subtrees of b except the two heaviest. Summing that up over the whole tree gives us a $O(n \log n)$ upper bound, as shown by the next lemma.   □

Lemma 6. *Let T be any unrooted tree of size n. For any vertex v let $S^v = \langle S_0^v, S_1^v, \ldots \rangle$ be the sequence of subtrees of v (i.e., the connected components of $T \setminus \{v\}$) ordered descending by their size, that is, $|V(S_i^v)| \geq |V(S_{i+1}^v)|$. Then for*

$$\Psi(v) = \sum_{i=2}^{|S^v|-1} |V(S_i^v)|$$

we have $\sum_{v \in V(T)} \Psi(v) \in O(n \log n)$.

Proof. Let r be a centroid point of T, that is, a vertex such that $|V(S_0^r)| \leq \frac{1}{2}|V(T)|$. We root T at r, and perform the heavy-light decomposition of T (see Definition 3). Observe that for all vertices $v \neq r$ we have that S_0^v contains r (it corresponds to the parent of v) and S_1^v corresponds to the biggest child of v. In other words, at most S_0^v and S_1^v can be connected by heavy edges, all the other subtrees $S_2^v, S_3^v, \ldots$ are connected by light edges.

Now we take an arbitrary vertex w and calculate how many times it can appear in $\sum_{v \in V(T)} \Psi(v)$. Suppose v is a vertex that counts w in $\Psi(v)$, then the first edge on the path from v to w has to be light, moreover, S_0^v is not counted in $\Psi(v)$, so that path cannot pass through the parent of v. Because of that v has to be an ancestor of w, however, there are at most $O(\log n)$ light edges on any path from w to the root r for any w. In other words, there can be at most $O(\log n)$ vertices that count w in its sum of Ψ. Summing that for all vertices of T we get the desired bound of $O(n \log n)$. □

We continue with the proof of Lemma 5, stated again below.

Lemma 5 The total length of final dispatching paths is $O(n \log^2 n)$, i.e.,

$$\sum_{\substack{t \in [n]: \\ b=\mathrm{dis}(\pi_t) \in B \\ t=\mathrm{tlast}(b)}} |\rho_t| \in O(n \log^2 n)$$

Proof. In order to bound the sum as claimed, we introduce some additional structure on T. We decompose T into paths which cover T. We pick an arbitrary vertex of T as a root. We adopt the heavy-light decomposition defined below.

Definition 3. *In the heavy-light decomposition each non-leaf node selects one* heavy *edge - the edge to the child that has the greatest number of descendants (breaking ties arbitrarily). The selected edges form the paths of the decomposition (called* heavy paths*). These heavy paths partition the vertices of T. Let* pheavy(v) *denote the heavy path containing v. A* light *edge is an edge of T that is not heavy.*

By construction, every path in T contains at most $O(\log n)$ light edges. In Fig. 1, the heavy paths in the tree are marked bold.

Now fix a black dispatching vertex b and the last time $t = \mathrm{tlast}(b)$ when b is dispatching. We bound the length of ρ_t by the length of λ_t, which is a path from b to a seed in S_{t-1}, that leaves each heavy path as soon as possible.

To be more precise, we define $\mathrm{closest}_t(v)$ as the closest vertex reachable from v at time t (in T_t^{int}), which belongs to pheavy(v) and has a light directed edge to an alive child at time t. Note that such a vertex exists if v is black and dispatching. Let light-child$_t(v)$ be the alive light child of v such that $v \to$ light-child$_t(v)$ at time t if such child exists. We now define a sequence of vertices:

$$\begin{cases} f_0 = b \\ e_i = \text{closest}_t(f_{i-1}) \text{ for } i = 1 \dots k \\ f_i = \text{light-child}_t(e_i) \text{ for } i = 1 \dots k \end{cases}$$

where k is the index when we reach a seed, i.e., either $e_i \in S_{t-1}$ or $f_i \in S_{t-1}$. We define $\lambda_t = f_0 \to e_1 \to f_1 \to \dots \to e_k/f_k$, see Fig. 1 for an illustration. Note, that λ_t is only defined for such t, that $t = \text{tlast}(b)$ for some black dispatching vertex b. We introduce a useful observation before we proceed.

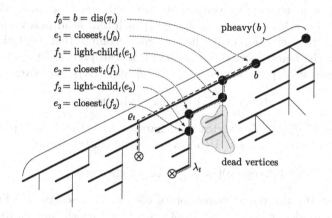

Fig. 1. The heavy-light decomposition and the definition of λ_t

Observation 3. *Any λ_t move towards the root only via heavy edges.*

As mentioned before, as ρ_t is the shortest path to a seed, we charge the cost of ρ_t onto the vertices of λ_t, which is certainly at least as long. The argument we are pursuing is going to be completed by the claim that every vertex is charged at most $O(\log^2 n)$ times during the runtime of SAP.

To that end we introduce the last definitions and observations. For any vertex v we define heavy-charge(v) to be the set of black dispatching vertices $b \in$ pheavy(v) such that at time $t = \text{tlast}(b)$ paths λ_t charge onto v:

$$\text{heavy-charge}(v) = \{b \in \text{pheavy}(v) \cap B : b = \text{dis}(\pi_t) \text{ and } t = \text{tlast}(b) \text{ and } v \in \lambda_t\}$$

We emphasize here that a black dispatching vertex $b = \text{dis}(\pi_t)$ of pheavy(v) can charge λ_t onto v at most once, and hence heavy-charge(v) is not a multiset.

Now fix a vertex w. We count how many times w is charged. Let charge(w) be the set of all black dispatching vertices that charge onto w the last time when they are dispatching:

$$\text{charge}(w) = \{b \in B : b = \text{dis}(\pi_t) \text{ and } t = \text{tlast}(b) \text{ and } w \in \lambda_t\}$$

Clearly, $|\text{charge}(w)|$ is the total number of times the vertex w is charged and that is what we want to bound. To complete the argument, we introduce one more definition.

Definition 4. *The head of a heavy path π, denoted as* head(π), *is the closest to the root vertex of π (closest in the undirected sense). A light ancestor of a vertex v, denoted as* light-ancestor(v), *is the parent in the tree T of the head of the heavy path containing v, i.e.,* light-ancestor(v) *is a parent of* head(pheavy(v)).

We now define a sequence of vertices starting with $w_0 = w$, such that $w_i = $ light-ancestor(w_{i-1}), for $i = 1 \ldots l$, where l is such that head(pheavy(w_l)) is the root of T. By the definition of the heavy-light decomposition, $l \in O(\log n)$. We observe that the black dispatching vertices that can potentially charge onto w are the vertices in $V(\text{pheavy}(w_0)) \cup \ldots \cup V(\text{pheavy}(w_l)))$. Moreover,

$$\text{charge}(w) \subseteq \bigcup_{i=0}^{l} \text{heavy-charge}(w_i)$$

since every black dispatching vertex that charges onto w that is in $V(\text{pheavy}(w_i))$ charges also onto w_i. Since sets $V(\text{pheavy}(w_i))$ are pairwise disjoint, this implies

$$|\text{charge}(w)| \leq \sum_{i=0}^{l} |\text{heavy-charge}(w_i)|$$

For the illustration of our construction of the charging scheme see Fig. 2. The black arrows mark the heavy charges of vertices w_i, which sum up to the total charge of w.

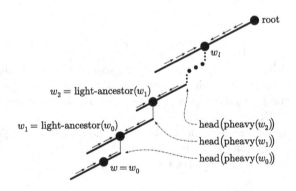

Fig. 2. The charging scheme

Below we prove Lemma 7, which states that for all $v \in V(T)$ it holds that $|\text{heavy-charge}(v)| \in O(\log(|\text{pheavy}(v)|))$. Having Lemma 7 at our disposal, we have $|\operatorname{charge}(w)| \le \sum_{i=0}^{l} |\operatorname{heavy-charge}(w_i)| \in O(\log^2 n)$. This completes the proof of Lemma 5. □

Lemma 7. *For all* $v \in V(T)$ *it holds that* $|\operatorname{heavy-charge}(v)| \in O(\log (|\operatorname{pheavy}(v)|))$.

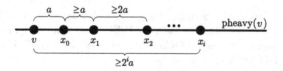

Fig. 3. The illustration to the proof of Lemma 7

Proof. We partition vertices of heavy-charge$(v) = X \cup Y \cup Z$ into pairwise disjoint sets X, Y, Z. We let $X = \{x_0 \dots x_p\}$ be the ancestors of v in heavy-charge(v) ordered in a way that x_{i+1} is the ancestor of x_i. Similarly, let $Y = \{y_0, \dots y_q\}$ be the descendants of v in heavy-charge(v) ordered in a way that y_{i+1} is a descendant of y_i. Finally we set $Z = \{v\}$ if v is a black dispatching vertex, otherwise $Z = \emptyset$. We focus on the number of vertices in X first. We use here $d(\bullet, \bullet)$ to denote the distance between two vertices in T in the undirected sense. Let $a = d(v, x_0)$. We prove inductively that $d(x_i, v) \ge 2^i a$, see Fig. 3 for an illustration. The claim clearly holds for $i = 0$. Now assume it holds for $j \le i$ for some i. Consider x_i and x_{i+1}. Let $t_i = \operatorname{tlast}(x_i)$ and $t_{i+1} = \operatorname{tlast}(x_{i+1})$. We distinguish two cases:

1. $t_i < t_{i+1}$. By definition of λ_{t_i}, it holds that $d(\operatorname{closest}_{t_i}(x_i), x_i) \ge d(x_i, v) \ge 2^i a$. Because x_i is dispatching at time t_i, there is at time t an alternative path λ'_{t_i} (going up the tree towards x_{i+1}) from x_i to a seed. Consider again two cases:

   (a) λ'_{t_i} does not cross x_{i+1}. This means that λ'_{t_i} leaves pheavy(v) in a vertex u that has a directed edge to a light alive child at time t_i. Hence,
   $$d(x_i, x_{i+1}) \ge d(x_i, u) \ge d(\operatorname{closest}_{t_i}(x_i), x_i) \ge 2^i a$$
   so $d(x_{i+1}, v) \ge 2^{i+1} a$.

   (b) λ'_{t_i} crosses x_{i+1}. Then, x_{i+1} at time t_i has a directed edge to an alive light child. This holds because if all reachable light children of x_{i+1} at time t_i are dead, then $\operatorname{degeff}_t(x_{i+1}) \le 1$ remains for $t \ge t_i$, so x_{i+1} cannot be dispatching at time t_{i+1}. So, since x_{i+1} does have a directed edge to an alive light child at time t_i, we get $d(x_{i+1}, x_i) \ge d(x_i, v)$ and thus $d(x_{i+1}, v) \ge 2^{i+1} a$.

2. $t_{i+1} < t_i$. By definition, $\lambda_{t_{i+1}}$ crosses x_i. By a similar argument as above, at time t_{i+1} vertex x_i has a directed edge to a light alive child. This is a contradiction, as in such case $\lambda_{t_{i+1}}$ leaves pheavy(v) in x_i.

The claim that we proved implies that $|X| \in O(\log | \text{pheavy}(v)|)$. We analogously show that $|Y| \in O(\log | \text{pheavy}(v)|)$. Since $|Z| \leq 1$, we obtain $| \text{heavy-charge}(v)| = |X \cup Y \cup Z| \in O(\log | \text{pheavy}(v)|)$. This completes the proof of Lemma 7 and the proof of Theorem 1. $\square$

References

1. Baswana, S., Gupta, M., Sen, S.: Fully dynamic maximal matching in $O(\log n)$ update time. In: Proceedings of the 2011 IEEE 52nd Annual Symposium on Foundations of Computer Science, FOCS 2011, ppp. 383–392. IEEE Computer Society, Washington, DC, USA (2011)
2. Bernstein, A., Stein, C.: Fully dynamic matching in bipartite graphs. In: Halldórsson, M.M., Iwama, K., Kobayashi, N., Speckmann, B. (eds.) ICALP 2015. LNCS, vol. 9134, pp. 167–179. Springer, Heidelberg (2015)
3. Bosek, B., Leniowski, D., Sankowski, P., Zych, A.: Online bipartite matching in offline time. In: 55th IEEE Annual Symposium on Foundations of Computer Science, FOCS 2014, Philadelphia, PA, USA, 18–21 October 2014, pp. 384–393. IEEE Computer Society (2014)
4. Chaudhuri, K., Daskalakis, C., Kleinberg, R.D., Lin, H.: Online bipartite perfect matching with augmentations. In: INFOCOM 2009, 28th IEEE International Conference on Computer Communications, Joint Conference of the IEEE Computer and Communications Societies, 19–25 April 2009, Rio de Janeiro, Brazil, pp. 1044–1052. IEEE (2009)
5. Edmonds, J., Karp, R.M.: Theoretical improvements in algorithmic efficiency for network flow problems. J. ACM **19**(2), 248–264 (1972)
6. Grove, E.F., Kao, M.Y., Krishnan, P., Vitter, J.S.: Online perfect matching and mobile computing. In: Akl, S.G., Dehne, F., Sack, J.-R., Santoro, N. (eds.) Algorithms and Data Structures. Lecture Notes in Computer Science, vol. 955, pp. 194–205. Springer, Heidelberg (1995)
7. Gupta, A., Kumar, A., Stein, C.: Maintaining assignments online: matching, scheduling, and flows. In: Chekuri, C., (ed.) Proceedings of the Twenty-Fifth Annual ACM-SIAM Symposium on Discrete Algorithms, SODA 2014, Portland, Oregon, USA, 5–7 January 2014, pp. 468–479. SIAM (2014)
8. Gupta, M., Peng, R.: Fully dynamic $(1+e)$-approximate matchings. In: IEEE 54th Annual Symposium on Foundations of Computer. Science, pp 548–557 (2013)
9. Ivković, Z., Lloyd, E.L.: Fully dynamic maintenance of vertex cover. In: Leeuwen, J. (ed.) Graph-Theoretic Concepts in Computer Science. Lecture Notes in Computer Science, vol. 790, pp. 99–111. Springer, Heidelberg (1994)
10. Karp, R.M., Upfal, E., Wigderson, A.: Constructing a perfect matching is in random NC. Combinatorica **6**(1), 35–48 (1986)
11. Mehta, A., Saberi, A., Vazirani, U., Vazirani, V.: Adwords and generalized on-line matching. In: 46th Annual IEEE Symposium on Foundations of Computer Science, FOCS 2005, pp. 264–273, October 2005
12. Neiman, O., Solomon, S.: Simple deterministic algorithms for fully dynamic maximal matching. In: Proceedings of the Forty-fifth Annual ACM Symposium on Theory of Computing, STOC 2013, pp. 745–754. ACM, New York, NY, USA (2013)
13. Onak, K., Rubinfeld, R.: Dynamic approximate vertex cover and maximum matching. In: Goldreich, O. (ed.) Property Testing, vol. 6390, pp. 341–345. Springer, Heidelberg (2010)

14. Sankowski, P.: Faster dynamic matchings and vertex connectivity. In: Proceedings of the Eighteenth Annual ACM-SIAM Symposium on Discrete Algorithms, SODA 2007, pp. 118–126. Society for Industrial and Applied Mathematics, Philadelphia, PA, USA (2007)

15. Sleator, D.D., Tarjan, R.E.: A data structure for dynamic trees. J. Comput. Syst. Sci. **26**(3), 362–391 (1983)

Buyback Problem with Discrete Concave Valuation Functions

Shun Fukuda[1], Akiyoshi Shioura[2][(✉)], and Takeshi Tokuyama[1]

[1] Graduate School of Information Sciences, Tohoku University,
Sendai 980-8579, Japan
{fukuda,tokuyama}@dais.is.tohoku.ac.jp
[2] Department of Social Engineering, Tokyo Institute of Technology,
Tokyo 152-8550, Japan
shioura.a.aa@m.titech.ac.jp

Abstract. We discuss an online discrete optimization problem called the buyback problem. In the literature of the buyback problem, the valuation function representing the value of a set of selected elements is given by a linear function. In this paper, we consider a generalization of the buyback problem using a nonlinear valuation function. We propose an online algorithm for the problem with a discrete concave valuation function, and show that it achieves the same competitive ratio as the best possible ratio for a linear valuation function.

1 Introduction

We discuss an online discrete optimization problem called the buyback problem. In the literature of the buyback problem, the valuation function representing the value of a set of elements is given by a linear (or additive) function. We refer to this variant of the buyback problem as the *linear buyback problem*. In this paper, we consider the *nonlinear buyback problem*, a generalization of the buyback problem with a nonlinear valuation function.

1.1 Model of Nonlinear Buyback Problem

To explain the nonlinear buyback problem, we consider a situation where a company wants to hire some workers from a set N of n applicants. Each applicant arrives one by one sequentially, and an interviewer of the company, which corresponds to an online algorithm, must decide immediately whether or not to hire the applicant. The company can hire at most $m > 0$ applicants; in addition, there may be some other constraints for a set of hired applicants due to their job skills and/or their human relationship. We denote by $\mathcal{F} \subseteq 2^N$ the set of feasible combinations of applicants. The interviewer wants to maximize the profit $v(X)$ obtained from a set $X \in \mathcal{F}$ of hired applicants. The function v is a nonlinear function in X in general since the job skill of applicants may overlap. It is natural to assume that function v is monotone nondecreasing and satisfies $v(\emptyset) = 0$

© Springer International Publishing Switzerland 2015
L. Sanità and M. Skutella (Eds.): WAOA 2015, LNCS 9499, pp. 72–83, 2015.
DOI: 10.1007/978-3-319-28684-6_7

and $v(X) > 0$ for $X \neq \emptyset$. It is often the case that a good applicant comes for an interview but addition of the applicant violates the feasibility. In such a case, the interviewer can add the applicant by canceling the contract with some previously hired applicant at the cost of some compensatory payment. In this paper, we assume that cancellation cost is given by a constant $c > 0$. It should be noted that applicants that are rejected at the interview or once accepted but canceled cannot be recovered later. The goal of the interviewer is to make an online decision to maximize the value $v(X)$ of hired applicants X, minus the total cancellation cost.

This online problem is called the buyback problem. More formally, the buyback problem is formulated as an online version of the following discrete optimization problem:

$$\text{Maximize } v(A \setminus C) - c|C| \qquad \text{subject to } C \subseteq A \subseteq N, \ A \setminus C \in \mathcal{F},$$

where A (resp., C) corresponds to a set of accepted (resp., once accepted but later canceled) elements, respectively. It is assumed that the set family $\mathcal{F}$ and the function v are accessible via appropriate oracles; that is, for a given set $X \subseteq N$, whether $X \in \mathcal{F}$ or not can be checked in constant time, and if $X \in \mathcal{F}$ then the function value $v(X)$ can be obtained in constant time.

For a special case of the buyback problem with a linear valuation function given as $v(X) = \sum_{i \in X} w(i)$ and a matroid constraint, Kawase, Han, and Makino [17] obtained the following result. It is assumed that a value $\ell > 0$ with $\ell \leq \min_{i \in N} w(i)$ is known in advance, and let

$$r^*(\ell, c) = 1 + \frac{c + \sqrt{c^2 + 4\ell c}}{2\ell}. \tag{1}$$

Note that the value $r^*(\ell, c)$ is dependent only on the ratio ℓ/c. For example, if $\ell/c = 2$ then $r^*(\ell, c) = 2$, and if $\ell/c = 6$ then $r^*(\ell, c) = 1.5$.

Theorem 1.1 ([17]). *Suppose that $v : 2^N \to \mathbb{R}$ is a linear valuation function and $\mathcal{F} \subseteq 2^N$ is the family of independent sets of a matroid. Then, the buyback problem admits an online algorithm with the competitive ratio $r^*(\ell, c)$. Moreover, there exists no online deterministic algorithm with a competitive ratio smaller than $r^*(\ell, c)$, even in the special case with $\mathcal{F} = \{X \subseteq N \mid |X| \leq 1\}$.*

The main aim of this paper is to generalize this result to the buyback problem with discrete concave valuation functions.

1.2 Our Result

In this paper, we present the first online algorithm for the nonlinear buyback problem and analyze its competitive ratio theoretically. Our main results given in Theorems 1.2 and 1.3 are proved in Sect. 3 by generalizing the approach used in [17] for the linear buyback problem. The analysis of competitive ratio in our setting, however, is much more difficult due to the nonlinearity of valuation function. We overcome this difficulty by utilizing discrete concavity of the function called M$^\natural$-concavity. M$^\natural$-concavity of the valuation function plays a crucial role in the analysis of competitive ratio of our online algorithm. It should be noted that while M$^\natural$-concave functions satisfy some kind of submodular inequality, submodularity alone is not enough to obtain the current result; see Concluding Remarks.

Buyback Problem with Gross Substitutes Valuations and Matching Weight Valuations. We first consider a nonlinear valuation function called a gross substitutes valuation. A valuation function $v : 2^N \to \mathbb{R}$ on 2^N is called a *gross substitutes valuation* (*GS valuation*, for short) if it satisfies the following condition:

$$\forall p, q \in \mathbb{R}^N \text{ with } p \le q, \forall X \in \arg\max_{U \subseteq N}\{v(U) - \sum_{i \in U} p(i)\},$$

$$\exists Y \in \arg\max_{U \subseteq N}\{v(U) - \sum_{i \in U} q(i)\} \text{ such that } \{i \in X \mid p(i)=q(i)\} \subseteq Y.$$

Intuitively, this condition is understood as follows, where N is regarded as a set of discrete items, and p and q are price vectors: if a buyer wants a set X of items at price p but some of the item prices are increased, then the buyer still wants items in X with unchanged prices (and possibly other items not in X).

A natural but nontrivial example of GS valuations arises from the maximum-weight matching problem on a complete bipartite graph, called *assignment valuations* [28] (or *OXS valuation* [20]). Going back to the situation at a company in Sect. 1.1, we suppose that the company has a set J of m jobs, to which hired workers are assigned. Each worker is assigned to at most one job in J, each job is assigned to at most one worker, and if worker $i \in N$ is assigned to a job $j \in J$, then profit $p(i, j) \in \mathbb{R}_{++}$ is obtained. Given a set $X \subseteq N$ of workers, the maximum total profit $v(X)$ obtained by assigning workers in X to jobs in J can be formulated as the maximum-weight matching problem on a complete bipartite graph G with the vertex sets N and J:

$$v(X) = \max \left\{ \sum_{(i,j) \in M} p(i, j) \,\middle|\, M : \text{matching in } G \text{ s.t. } \partial_N M = X \right\}, \quad (2)$$

where $\partial_N M$ denotes the set of vertices in N covered by edges in M. It is known that this function $v : 2^N \to \mathbb{R}$ is a GS valuation function [20,28].

The concept of GS valuation is introduced in Kelso and Crawford [18], where the existence of a Walrasian equilibrium is shown in a fairly general two-sided matching model. Since then, this concept plays a central role in mathematical economics and in auction theory, and is widely used in various economic models (see, e.g., [5,6,11–13,20]). The class of GS valuations is a proper subclass of submodular functions, and includes natural classes of valuations such as weighted rank functions of matroids [7,9] and laminar concave function [23] (or *S-valuation* [5]), in addition to assignment valuations explained above. While GS valuation is a sufficient condition for the existence of a Walrasian equilibrium [18], it is also a necessary condition in some sense [13]. GS valuation is also related to desirable properties in the auction design [6,11,20]. See also [26,30] for more details on GS valuations as well as other related concepts.

We propose an online algorithm for the nonlinear buyback problem with a GS valuation and a cardinality constraint. We assume that a positive real number ℓ satisfying

$$\ell \leq \min\{v(X)/|X| \mid \emptyset \neq X \in \mathcal{F}\} \tag{3}$$

is known in advance. Note that this condition is a natural generalization of the condition used in [17]; indeed, for a linear valuation function, condition (3) is equivalent to $\ell \leq \min_{i \in N} w(i)$. In addition, if v is an assignment valuation function in (2), then every ℓ with $\ell \leq \min\{p(i,j) \mid i \in N, j \in J\}$ satisfies (3).

Theorem 1.2. *For a gross substitutes valuation function $v : 2^N \to \mathbb{R}$ and a cardinality constraint $\mathcal{F} = \{X \subseteq N \mid |X| \leq m\}$, the nonlinear buyback problem admits an online algorithm with the competitive ratio $r^*(\ell, c)$ in (1).*

It should be noted that our online algorithm does not require the information about the number of elements in N and the integer m.

Buyback Problem with Discrete Concave Valuations. Moreover, we consider a more general setting where $\mathcal{F}$ is a matroid and valuation function $v : \mathcal{F} \to \mathbb{R}$ is a discrete concave function called $M^\natural$-*concave function*. It is known that a family $\mathcal{F} \subseteq 2^N$ of matroid independent sets satisfies the following property [25]:

($\mathbf{B}^\natural$ – **EXC**) $\forall X, Y \in \mathcal{F}, \forall i \in X \setminus Y$, at least one of (i) and (ii) holds:
(i) $X - i \in \mathcal{F}$, $Y + i \in \mathcal{F}$, (ii) $\exists j \in Y \setminus X \colon X - i + j \in \mathcal{F}$, $Y + i - j \in \mathcal{F}$,

where $X - i + j$ is a short-hand notation for $(X \setminus \{i\}) \cup \{j\}$. We consider a function $v : \mathcal{F} \to \mathbb{R}$ defined on matroid independent sets $\mathcal{F}$. A function v is said to be $M^\natural$-*concave* [25] (read "M-natural-concave") if it satisfies the following:

($\mathbf{M}^\natural$ – **EXC**) $\forall X, Y \in \mathcal{F}, \forall i \in X \setminus Y$, at least one of (i) and (ii) holds:
(i) $X - i \in \mathcal{F}$, $Y + i \in \mathcal{F}$, and $v(X) + v(Y) \leq v(X - i) + v(Y + i)$,
(ii) $\exists j \in Y \setminus X \colon X - i + j \in \mathcal{F}$, $Y + i - j \in \mathcal{F}$,
and $v(X) + v(Y) \leq v(X - i + j) + v(Y + i - j)$.

The concept of $M^\natural$-concave function is introduced by Murota and Shioura [25] (independently of GS valuations) as a class of discrete concave functions. $M^\natural$-concavity is originally introduced for functions defined on integer lattice points (see, e.g., [23]), and the present definition of $M^\natural$-concavity for set functions can be obtained by specializing the original definition through the one-to-one correspondence between set functions and functions defined on $\{0, 1\}$-vectors. The concept of $M^\natural$-concave function is an extension of the concept of M-concave function introduced by Murota [21,22]. The concepts of $M^\natural$-concavity/M-concavity play primary roles in the theory of discrete convex analysis [23], which provides a framework for tractable nonlinear discrete optimization problems.

M$^\natural$-concave functions have various desirable properties as discrete concavity. Global optimality is characterized by local optimality, which implies the validity of a greedy algorithm for M$^\natural$-concave function maximization. Maximization of an M$^\natural$-concave function can be done efficiently in polynomial time (see, e.g., [23, 25]).

The class of M$^\natural$-concave functions includes linear functions on matroids. Hence, the M$^\natural$-concave buyback problem (i.e., the buyback problem with an M$^\natural$-concave valuation) is a proper generalization of the linear buyback problem with a matroid constraint discussed in Kawase et al. [17]. Furthermore, the M$^\natural$-concave buyback problem also includes the problem with a GS valuation function and a cardinality constraint as a special case.

In this paper, we show the following result for the M$^\natural$-concave buyback problem.

Theorem 1.3. *If $\mathcal{F} \subseteq 2^N$ is the family of independent sets of a matroid and $v : \mathcal{F} \to \mathbb{R}$ is an M$^\natural$-concave function, then the nonlinear buyback problem admits an online algorithm with the competitive ratio $r^*(\ell, c)$ in (1).*

This theorem implies Theorem 1.2 as a corollary. In addition, this theorem also implies the former statement of Theorem 1.1, hence generalizing the result of Kawase et al. [17]. The latter statement in Theorem 1.1 shows that our competitive ratio in Theorem 1.3 is the best possible for the M$^\natural$-concave buyback problem.

1.3   Related Work

We review some previous results on the linear buyback problem and some related results. In the literature of the linear buyback problem, two types of cancellation cost are considered so far: *proportional* cost and *unit* cost; the latter one is used in this paper. In the case of proportional cost, we are given a constant $f > 0$ and the cancellation cost of each element u is equal to $fw(u)$ if $w(u)$ is the value of u. In the case of unit cost, we are given a constant $c > 0$ and the cancellation cost of each element u is equal to c. Note that in the nonlinear buyback problem, unit cancellation cost is more suitable since proportional cancellation cost is heavily dependent on the linearity of a valuation function.

The linear buyback problem is originally modeled by using proportional cost. In this setting, Babaioff et al. [3] and Constantin et al. [10] independently proposed deterministic online algorithms for the problem with single matroid constraint, where the competitive ratio is $1 + 2f + 2\sqrt{f(1 + f)}$. Babaioff et al. [4] also showed that this competitive ratio is the best possible bound for deterministic algorithms, and presented a randomized algorithm with a better competitive ratio in the case of small f. Later, Ashwinkumar and Kleinberg [2] proposed a randomized algorithm with an improved competitive ratio, which is shown to be the best possible. Ashwinkumar [1] considered a more general constraints such as the intersection of multiple matroids, and proposed online algorithms with theoretical bounds for the competitive ratio. Some variants of knapsack constraint were also considered in [3, 4, 14].

The linear buyback problem with unit cost was first introduced by Han et al. [14]. Some variants of knapsack constraints are considered in [14, 17], while single matroid constraint is considered by Kawase et al. [17] (see Theorem 1.1).

Variants of the buyback problem with zero cancellation cost are also extensively discussed in the literature. One such example is the problem under a knapsack constraint, which is referred to as the *online removal knapsack problem* (see, e.g., [15, 16]). Recently, the nonlinear buyback problem with zero cancellation cost and submodular valuation function (called the *online submodular maximization with preemption*) is considered by Buchbinder et al. [8]. Note that the linear buyback problem with a single matroid constraint is trivial if the cancellation cost is zero; indeed, existing online algorithms for this problem reduce to variants of greedy algorithms that find an (offline) optimal solutions.

The buyback problem with an assignment valuation function can be seen as a variant of online bipartite matching problems, where vertices on the one side of a bipartite graph (corresponding to applicants) arrive online one by one (see, e.g., [19] and the references therein). Among many variants of such online matching problems, our problem setting is different in the following two points. First, we allow re-assignment of previously accepted vertices to the vertices on the other side whenever a newly arrived vertex is accepted. Second, we allow exchange of a previously accepted vertex with a newly arrived vertex by paying a cancellation cost. Without a cancellation cost, our online matching problem is trivial since we allow re-assignment; indeed, it is easy to construct an online algorithm that finds an (offline) optimal matching under this setting.

2   M♮-concave Functions and GS Valuations

In this section we review the concept of M♮-concavity and its connection with GS valuation.

Let $\mathcal{F}$ be the family of independent sets of a matroid. A function $v : \mathcal{F} \to \mathbb{R}$ is said to be *M♮-concave* if it satisfies the condition (M♮-EXC). It is known that every M♮-concave function is a submodular function in the following sense (cf. [23]):

Proposition 2.1 ([23, Theorem 6.19]). *Let $f : \mathcal{F} \to \mathbb{R}$ be an M♮-concave function defined on a family $\mathcal{F} \subseteq 2^N$ of matroid independent sets. For $X, Y \in \mathcal{F}$ with $X \cup Y \in \mathcal{F}$, it holds that $v(X) + v(Y) \geq v(X \cup Y) + v(X \cap Y)$.*

From the condition (M♮-EXC) we can obtain the following property.

Proposition 2.2 ([25, Theorem 4.2]). *Let $f : \mathcal{F} \to \mathbb{R}$ be an M♮-concave function defined on matroid independent sets $\mathcal{F}$. For every $X, Y \in \mathcal{F}$ with $|X| = |Y|$ and $u \in X \setminus Y$, there exists some $v \in Y \setminus X$ such that $f(X) + f(Y) \leq f(X - u + v) + f(Y + u - v)$.*

Note that the sum of an M♮-concave function and a linear function is again an M♮-concave function, while the sum of two M♮-concave functions is not M♮-concave in general.

The next property shows the connection between M$^\natural$-concavity and gross substitute valuation. In particular, the property below shows that the buyback problem with a gross substitute valuation function is a special case of M$^\natural$-concave buyback problem.

Theorem 2.1 (*cf.* [12]). *Let $v : 2^N \to \mathbb{R}$ be a function defined on 2^N.*
(i) *v is a GS valuation function if and only if it is $M^\natural$-concave.*
(ii) *Suppose that v is a GS valuation function and let m be a nonnegative integer. Then, the function $v_m : \mathcal{F}_m \to \mathbb{R}$ given by $\mathcal{F}_m = \{X \in 2^N \mid |X| \le m\}$ and $v_m(X) = v(X)$ ($X \in \mathcal{F}_m$) is an $M^\natural$-concave function.*

A simple example of M$^\natural$-concave function is a linear function $f(X) = w(X)$ ($X \in \mathcal{F}$) defined on a family $\mathcal{F} \subseteq 2^N$ of matroid independent sets, where $w \in \mathbb{R}^N$. In particular, if $\mathcal{F} = 2^N$ then f is a GS valuation function. Below we give some nontrivial examples of M$^\natural$-concave functions and GS valuation functions. See [23, 24] for more examples.

Example 2.1 (Maximum-weight bipartite matching). In Sect. 1.2 we explained an assignment valuation as an example of GS valuations, where a complete bipartite graph is used. By using a non-complete bipartite graph instead, we can obtain an example of M$^\natural$-concave functions as follows.

Consider a bipartite graph G with two vertex sets N, J and an edge set E ($\subseteq N \times J$), where N and J correspond to workers and jobs, respectively. An edge $(i, j) \in E$ means that worker $i \in N$ has ability to process job $j \in J$, and profit $p(i, j) \in \mathbb{R}_{++}$ can be obtained by assigning worker i to job j. Consider a matching between workers and jobs, and define $\mathcal{F} \subseteq 2^N$ by

$$\mathcal{F} = \{X \subseteq N \mid \exists M : \text{matching in } G \text{ s.t. } \partial_N M = X\}.$$

It is well known that $\mathcal{F}$ is a family of independent sets in a transversal matroid (see, e.g., [27]). Define $v : \mathcal{F} \to \mathbb{R}$ by

$$v(X) = \max\Big\{ \sum_{(i,j) \in M} p(i,j) \mid M : \text{matching in } G \text{ s.t. } \partial_N M = X \Big\} \quad (X \in \mathcal{F}).$$

Then, v is an M$^\natural$-concave function [24, Sect. 11.4.2]. □

Example 2.2 (Laminar concave functions). Let $\mathcal{T} \subseteq 2^N$ be a laminar family, i.e., $X \cap Y = \emptyset$ or $X \subseteq Y$ or $X \supseteq Y$ holds for every $X, Y \in \mathcal{T}$. For $Y \in \mathcal{T}$, let $\varphi_Y : \mathbb{Z}_+ \to \mathbb{R}$ be a univariate concave function. Define a function $v : 2^N \to \mathbb{R}$ by

$$v(X) = \sum_{Y \in \mathcal{T}} \varphi_Y(|X \cap Y|) \quad (X \in 2^N),$$

which is called a *laminar concave function* [23, Sect. 6.3] (also called an *S-valuation* in [5]). Special cases of laminar concave functions are a *downward sloping symmetric function* [11] given as $v(X) = \varphi(|X|)$ and a *nested concave function* given as

$$v(X) = \sum_{i=1}^{n} \varphi_i(|X \cap \{1, 2, \ldots, i\}|),$$

where φ and φ_i $(i \in N)$ are univariate concave functions. Every laminar concave function is a GS valuation function. □

Example 2.3 (Weighted rank functions). Let $\mathcal{I} \subseteq 2^N$ be the family of independent sets of a matroid, and $w \in \mathbb{R}_+^N$. Define a function $v : 2^N \to \mathbb{R}_+$ by

$$v(X) = \max\{w(Y) \mid Y \subseteq X, \ Y \in \mathcal{I}\} \quad (X \in 2^N),$$

which is called the *weighted rank function* [9]. If $w(i) = 1$ $(i \in N)$, then v is an ordinary rank function of the matroid $(N, \mathcal{I})$. Every weighted rank function is a GS valuation function [29]. □

3 Our Online Algorithm and Analysis

In this section, we propose an online algorithm for $M^\natural$-concave buyback problem and analyze its competitive ratio.

3.1 Algorithm

Recall that the cancellation cost c and the value ℓ satisfying (3) is known in advance. We assume that $N = \{i_1, i_2, \ldots, i_n\}$ and the elements in N arrive in this order. In each iteration, the algorithm maintains a set $B_k \in \mathcal{F}$. To control the number of cancellations, we use an increasing sequence of real numbers $\psi(t)$ $(t = 1, 2, \ldots)$ as parameters, which will be determined later by using c and ℓ. We assume that $\psi(1) = 0$ and $\psi(t+1) - \psi(t)$ is nondecreasing with respect to t.

In the k-th iteration, the algorithm adds an element i_k (i.e., set $B_k = B_{k-1} + i_k$) if $B_{k-1} + i_k \in \mathcal{F}$ and $v(B_{k-1} + i_k) > v(B_{k-1})$. Otherwise, the algorithm tries to exchange an element j_k in B_{k-1} satisfying $B_{k-1} - j_k + i_k \in \mathcal{F}$ and

$$v(B_{k-1} - j_k + i_k) = \max\{v(B_{k-1} - j + i_k) \mid j \in B_{k-1}, \ B_{k-1} - j + i_k \in \mathcal{F}\}. \quad (4)$$

If the value $v(B_{k-1} - j_k + i_k)$ is large enough compared to $v(B_{k-1})$, then the algorithm replace j_k with i_k; otherwise, the algorithm does not add and sets $B_k = B_{k-1}$. A detailed description of the algorithm is as follows.

Algorithm $M^\natural$BP
Step 0: Set $B_0 = \emptyset$.
Step 1: For each element i_k, $k = 1, 2, \ldots, n$, in order of arrival, do the following:
[Case 1: $B_{k-1} + i_k \in \mathcal{F}$] Set $B_k = B_{k-1} + i_k$.
[Case 2: $B_{k-1} + i_k \notin \mathcal{F}$] Let $j_k \in B_{k-1}$ be an element satisfying (4). If $v(B_{k-1} - j_k + i_k) \geq \psi(t) + \ell \cdot |B_{k-1}| > v(B_{k-1})$ for some t, then set $B_k = B_{k-1} - j_k + i_k$ ("cancel j_k"); otherwise, set $B_k = B_{k-1}$ ("reject i_k").
Step 2: Output B_n. □

3.2 Bounding the Optimal Value

Let $B^* \in \mathcal{F}$ be an (offline) optimal solution of $M^\natural$-concave buyback problem. That is, $B^* \in \arg\max\{v(B) \mid B \in \mathcal{F}\}$. To analyze the competitive ratio of the algorithm above, we need to bound the value of $v(B^*)$ from above.

For $k = 1, 2, \ldots, n$, let t_k be the integer with $v(B_k) - \ell \cdot |B_k| \in [\psi(t_k), \psi(t_k + 1))$. We will derive the following upper bound of $v(B^*)$.

Lemma 3.1. $v(B^*) \leq v(B_n) + m(\psi(t_n + 1) - \psi(t_n))$.

To prove Lemma 3.1, we first show that the value $v(B^*)$ can be bounded from above in terms of the output B_n of the algorithm.

For two sets $B, B' \in \mathcal{F}$ with $|B| = |B'|$, we define $G(B, B')$, called the *exchangeability graph*, as a bipartite graph having $(B \setminus B', B' \setminus B)$ as the vertex bipartition and

$$\{(j, i) \mid j \in B \setminus B', \ i \in B' \setminus B, \ B - j + i \in \mathcal{F}\}$$

as the edge set. Note that $|B \setminus B'| = |B' \setminus B|$ holds since B and B' have the same cardinality, and $G(B, B')$ has a perfect matching (see, e.g., [27, Corollary 39.12a]).

For each edge (j, i) in $G(B, B')$, we define the weight of (j, i) by $v(B, j, i)$ given by

$$v(B, j, i) = v(B - j + i) - v(B).$$

Denote by $\widehat{v}(B, B')$ the maximum weight of a perfect matching in $G(B, B')$ with respect to the edge weight $v(B, j, i)$. We can bound the value $v(B')$ from above by using $v(B)$ and $\widehat{v}(B, B')$ as follows.

Lemma 3.2 [cf. [21, Lemma 3.4]]. For $B, B' \in \mathcal{F}$ with $|B| = |B'|$, it holds that $v(B') \leq v(B) + \widehat{v}(B, B')$.

We denote $m = \max\{|X| \mid X \in \mathcal{F}\}$. Note that $|B_n| = |B^*| = m$ holds since $\mathcal{F}$ is a family of matroid independent sets and v is monotone nondecreasing. Hence, the following inequality follows immediately from Lemma 3.2.

Lemma 3.3. $v(B^*) \leq v(B_n) + \sum_{i \in B^* \setminus B_n} \max\{v(B_n, j, i) \mid j \in B_n\}$.

To bound the value $\max\{v(B_n, j, i) \mid j \in B_n\}$ in Lemma 3.3, we show a useful inequality for the value $v(B_k, j, i)$, which plays a key role in the analysis. For $k = 1, 2, \ldots, n$, let

$$C_k = \{j_t \mid j_t \text{ is canceled in Case 2 of the } h\text{-th iteration}, 1 \leq h \leq k\},$$
$$R_k = \{i_t \mid i_t \text{ is rejected in Case 2 of the } h\text{-th iteration}, 1 \leq h \leq k\}.$$

Note that the sets B_k, C_k, and R_k provide a partition of set $\{1, 2, \ldots, k\}$.

Lemma 3.4. For $k = 1, 2, \ldots, n$, $j \in B_k$, and $i \in C_k \cup R_k$, it holds that

$$v(B_k, j, i) \leq \begin{cases} 0 & (\text{if } i \in C_k), \\ \max\{v(B_{h-1}, j', i_h) \mid j' \in B_{h-1}\} & (\text{if } i = i_h \in R_k \text{ with } h \leq k). \end{cases}$$
$$(5)$$

Using Lemma 3.4, we get a bound for $\max\{v(B_n, j, i) \mid j \in B_n\}$.

Lemma 3.5. *For* $i \in N \setminus B_n$, $\max\{v(B_n, j, i) \mid j \in B_n\} \le \psi(t_n + 1) - \psi(t_n)$.

Lemma 3.1 follows immediately from Lemmas 3.3 and 3.5.

3.3 Analysis of Competitive Ratio

We now prove that our online algorithm achieves the competitive ratio $r^*(\ell, c)$ in (1) by setting values $\psi(t)$ $(t = 1, 2, \ldots)$ appropriately.

We consider the set of intervals given by values $\psi(t)$, and denote the length of the t-th interval as $\lambda(t) = \psi(t + 1) - \psi(t)$. Note that whenever some element is canceled in some iteration of our algorithm, the value $v(B_k) - \ell|B_k|$ moves to some upper interval. Since $v(B_n) - \ell m \in [\psi(t_n), \psi(t_n + 1))$, our algorithm cancels some elements at most $t_n - 1$ times, and therefore the payoff obtained by the algorithm is at least $v(B_n) - (t_n - 1)c$. By this fact and Lemma 3.1, the competitive ratio of the algorithm is at most

$$\frac{v(B^*)}{v(B_n) - (t_n - 1)c} \le \frac{v(B_n) + m\lambda(t_n)}{v(B_n) - (t_n - 1)c} \le \frac{(\psi(t_n) + \ell m) + m\lambda(t_n)}{(\psi(t_n) + \ell m) - (t_n - 1)c}$$

$$\le \max_{t \ge 1} \frac{(\psi(t) + \ell m) + m\lambda(t)}{(\psi(t) + \ell m) - (t - 1)c}, \quad (6)$$

where the second inequality follows from the inequality $\psi(t_n) + \ell m \le v(B_n)$ and the fact that for $p, q \in \mathbb{R}_+$ the function $(x + p)/(x - q)$ in x is nonincreasing in the interval $(q, +\infty)$. We denote by r the ratio in the last term of (6). In the following, we analyze the minimum value r of the ratio. Note that $r > 1$.

We will set values $\psi(t)$ so that

$$\frac{(\psi(t) + \ell m) + m\lambda(t)}{(\psi(t) + \ell m) - (t - 1)c} = \frac{(\psi(t) + \ell m) + m(\psi(t + 1) - \psi(t))}{(\psi(t) + \ell m) - (t - 1)c} = r$$

holds for all $t \ge 1$. This implies the following recursive formula for $\psi(t)$:

$$\psi(1) = 0, \qquad \psi(t + 1) = \frac{m - 1 + r}{m}(\psi(t) + \ell m) - \frac{cr}{m}(t - 1) + \ell. \quad (7)$$

By solving this recursive formula, we have $r = 1 + \frac{c + \sqrt{c^2 + 4\ell c}}{2\ell} = r^*(\ell, c)$, i.e., the competitive ratio of our algorithm is $r^*(\ell, c)$. This concludes the proof of Theorem 1.3.

4 Concluding Remarks

We have shown that the competitive ratio of our online algorithm for $M^\natural$-concave buyback problem is $r^*(\ell, c)$. Note that $r^*(\ell, 0) = 1$, which means that our online algorithm finds an offline optimal solution by setting $c = 0$.

It should be noted that our approach does not extend to the nonlinear buy-back problem with a submodular valuation function. To illustrate this, let us consider an instance of the buyback problem, where $N = \{i_1, i_2, i_3, i_4\}$, the valuation function $v : 2^N \to \mathbb{R}$ is given by

$$v(\emptyset) = 0, \quad v(\{i_1\}) = v(\{i_2\}) = 2, \quad v(\{i_3\}) = v(\{i_4\}) = 3,$$
$$v(X) = 6 \text{ if } |X| \geq 2 \text{ and } X \supseteq \{i_3, i_4\}$$
$$v(X) = 4 \text{ if } |X| = 2 \text{ and } X \neq \{i_3, i_4\},$$
$$v(N \setminus \{i_3\}) = v(N \setminus \{i_4\}) = 5,$$

and the constraint is $\mathcal{F} = \{X \in 2^N \mid |X| \leq 2\}$. It can be checked that the function v is submodular but not $M^\natural$-concave.

Suppose that our online algorithm is applied to this instance, where the elements i_1, i_2, i_3, i_4 arrive in this order. Then, the algorithm first accepts elements i_1 and i_2, and then rejects i_3 and i_4 since the function value cannot be increased by swapping new elements with old elements one by one. Hence, the value of the output is $v(\{i_1, i_2\}) = 4$. Note that this behavior of the algorithm is irrelevant to the choice of the cancellation cost c. On the other hand, an offline optimal solution is $B^* = \{i_3, i_4\}$, for which $v(B^*) = 6$. Hence, the competitive ratio of our algorithm is at least $6/4 = 1.5$, while the ratio $r^*(\ell, c)$ can be close to 1 if we choose a sufficiently small positive c. This fact shows that our algorithm and analysis in this paper do not extend to submodular valuation functions.

Acknowledgements. The authors thank anonymous referees for their valuable comments on the manuscript. This work is supported by JSPS/MEXT KAKENHI Grand Numbers 24106007, 25106503, 15H02665, 15H00848, 15K00030.

References

1. Ashwinkumar, B.V.: Buyback problem - approximate matroid intersection with cancellation costs. In: Aceto, L., Henzinger, M., Sgall, J. (eds.) ICALP 2011, Part I. LNCS, vol. 6755, pp. 379–390. Springer, Heidelberg (2011)
2. Ashwinkumar, B.V., Kleinberg, R.: Randomized online algorithms for the buyback problem. In: Leonardi, S. (ed.) WINE 2009. LNCS, vol. 5929, pp. 529–536. Springer, Heidelberg (2009)
3. Babaioff, M., Hartline, J.D., Kleinberg, R.D.: Selling banner ads: online algorithms with buyback. In: Proceedings of 4th Workshop on Ad Auctions (2008)
4. Babaioff, M., Hartline, J.D., Kleinberg, R.D.: Selling ad campaigns: online algorithms with cancellations. In: EC 2009, pp. 61–70. ACM, New York (2009)
5. Bing, M., Lehmann, D., Milgrom P.: Presentation and structure of substitutes valuations. In: EC 2004, pp. 238–239. ACM, New York (2004)
6. Blumrosen, L., Nisan, N.: Combinatorial auction. In: Nisan, N., Roughgarden, T., Tardos, É., Vazirani, V.V. (eds.) Algorithmic Game Theory, pp. 267–299. Cambridge University Press, New York (2007)
7. Buchbinder, N., Naor, J.S., Ravi, R., Singh, M.: Approximation algorithms for online weighted rank function maximization under matroid constraints. In: Czumaj, A., Mehlhorn, K., Pitts, A., Wattenhofer, R. (eds.) ICALP 2012, Part I. LNCS, vol. 7391, pp. 145–156. Springer, Heidelberg (2012)

8. Buchbinder, N., Feldman, M., Schwartz, R.: Online submodular maximization with preemption. SODA **2015**, 1202–1216 (2015)
9. Calinescu, G., Chekuri, C., Pál, M., Vondrák, J.: Maximizing a submodular set function subject to a matroid constraint. SIAM J. Comput. **40**, 1740–1766 (2011)
10. Constantin, F., Feldman, J., Muthukrishnan, S., Pál, M.: An online mechanism for ad slot reservations with cancellations. SODA **2009**, 1265–1274 (2009)
11. Cramton, P., Shoham, Y., Steinberg, R.: Combinatorial Auctions. MIT Press, Cambridge (2006)
12. Fujishige, S., Yang, Z.: A note on Kelso and Crawford's gross substitutes condition. Math. Oper. Res. **28**, 463–469 (2003)
13. Gul, F., Stacchetti, E.: Walrasian equilibrium with gross substitutes. J. Econ. Theor. **87**, 95–124 (1999)
14. Han, X., Kawase, Y., Makino, K.: Online knapsack problem with removal cost. Algorithmica **70**, 76–91 (2014)
15. Han, X., Kawase, Y., Makino, K.: Randomized algorithms for removable online knapsack problems. In: Fellows, M., Tan, X., Zhu, B. (eds.) FAW-AAIM 2013. LNCS, vol. 7924, pp. 60–71. Springer, Heidelberg (2013)
16. Iwama, K., Taketomi, S.: Removable online knapsack problems. In: Widmayer, P., Triguero, F., Morales, R., Hennessy, M., Eidenbenz, S., Conejo, R. (eds.) ICALP 2002. LNCS, vol. 2380, pp. 293–305. Springer, Heidelberg (2002)
17. Kawase, Y., Han, X., Makino, K.: Unit cost buyback problem. In: Cai, L., Cheng, S.-W., Lam, T.-W. (eds.) Algorithms and Computation. LNCS, vol. 8283, pp. 435–445. Springer, Heidelberg (2013)
18. Kelso Jr., A.S., Crawford, V.P.: Job matching, coalition formation and gross substitutes. Econometrica **50**, 1483–1504 (1982)
19. Kesselheim, T., Radke, K., Tönnis, A., Vöcking, B.: An optimal online algorithm for weighted bipartite matching and extensions to combinatorial auctions. In: Bodlaender, H.L., Italiano, G.F. (eds.) ESA 2013. LNCS, vol. 8125, pp. 589–600. Springer, Heidelberg (2013)
20. Lehmann, B., Lehmann, D., Nisan, N.: Combinatorial auctions with decreasing marginal utilities. Games Econom. Behav. **55**, 270–296 (2006)
21. Murota, K.: Valuated matroid intersection I: optimality criteria. SIAM J. Discrete Math. **9**, 545–561 (1996)
22. Murota, K.: Discrete convex analysis. Math. Program. **83**, 313–371 (1998)
23. Murota, K.: Discrete Convex Analysis. SIAM, Philadelphia (2003)
24. Murota, K.: Recent developments in discrete convex analysis. In: Cook, W.J., Lovász, L., Vygen, J. (eds.) Research Trends in Combinatorial Optimization, pp. 219–260. Springer, Berlin (2009)
25. Murota, K., Shioura, A.: M-convex function on generalized polymatroid. Math. Oper. Res. **24**, 95–105 (1999)
26. Paes Leme, R.: Gross substitutability: An algorithmic survey. preprint (2014)
27. Schrijver, A.: Combinatorial Optimization: Polyhedra and Efficiency. Springer, Berlin (2003)
28. Shapley, L.: Complements and substitutes in the optimal assignment problem. Naval Res. Logist. Quart. **9**, 45–48 (1962)
29. Shioura, A.: On the pipage rounding algorithm for submodular function maximization: a view from discrete convex analysis. Disc. Math. Alg. Appl. **1**, 1–23 (2009)
30. Shioura, A., Tamura, A.: Gross substitutes condition and discrete concavity for multi-unit valuations: a survey. J. Oper. Res. Soc. Jpn. **58**, 61–103 (2015)

On Temporally Connected Graphs
of Small Cost

Eleni C. Akrida[1]([✉]), Leszek Gąsieniec[1], George B. Mertzios[2],
and Paul G. Spirakis[1]

[1] Department of Computer Science, University of Liverpool, Liverpool, UK
{Eleni.Akrida,L.A.Gasieniec,P.Spirakis}@liverpool.ac.uk
[2] School of Engineering and Computing Sciences, Durham University, Durham, UK
George.Mertzios@durham.ac.uk

Abstract. We study the design of small cost temporally connected
graphs, under various constraints. We mainly consider undirected graphs
of n vertices, where each edge has an associated set of discrete availability instances (labels). A journey from vertex u to vertex v is a path
from u to v where successive path edges have strictly increasing labels.
A graph is temporally connected iff there is a (u, v)-journey for any pair
of vertices u, v, $u \neq v$. We first give a simple polynomial-time algorithm
to check whether a given temporal graph is temporally connected. We
then consider the case in which a designer of temporal graphs can *freely
choose* availability instances for all edges and aims for temporal connectivity with very small *cost*; the cost is the total number of availability
instances used. We achieve this via a simple polynomial-time procedure
which derives designs of cost linear in n, and at most the optimal cost
plus 2. To show this, we prove a lower bound on the cost for any undirected graph. However, there are pragmatic cases where one is not free
to design a temporally connected graph anew, but is instead *given* a
temporal graph design with the claim that it is temporally connected,
and wishes to make it more cost-efficient by removing labels without
destroying temporal connectivity (redundant labels). Our main technical result is that computing the maximum number of redundant labels
is APX-hard, i.e., there is no PTAS unless $P = NP$. On the positive
side, we show that in dense graphs with random edge availabilities, all
but $\Theta(n)$ labels are redundant whp. A temporal design may, however,
be *minimal*, i.e., no redundant labels exist. We show the existence of
minimal temporal designs with at least $n \log n$ labels.

1 Introduction and Motivation

A temporal network is, roughly speaking, a network that changes with time.
A great variety of modern and traditional networks are dynamic, e.g., social

Supported in part by (i) the School of EEE and CS and the NeST initiative of the
University of Liverpool, (ii) the FET EU IP Project MULTIPLEX under contract
No. 317532, and (ii) the EPSRC Grant EP/K022660/1.

L. Sanità and M. Skutella (Eds.): WAOA 2015, LNCS 9499, pp. 84–96, 2015.
DOI: 10.1007/978-3-319-28684-6_8

networks, wireless networks, transport networks. Dynamic networks have been attracting attention over the past years [3,4,7,9,21], exactly because they model real-life applications. Following the model of [1,13,20], we consider *discrete time* and restrict attention to systems in which the connections between the participating entities may change but the entities remain unchanged. This assumption is clearly natural when the dynamicity of the system is inherently discrete and gives a purely combinatorial flavor to the resulting models and problems.

In several such dynamic settings, maintaining connections may come at a cost; consider a transport network or an unstable chemical or physical structure, where energy is required to keep a link available. We define the cost as the total number of discrete time instances, e.g., days or hours, at which the network links become available, i.e., the sum over all edges of the number of the edge's availability instances. We focus on design issues of temporal networks that are temporally connected; a temporal network is temporally connected if information can travel over time from any node to any other node following *journeys*, i.e., paths whose successive edges have strictly increasing availability time instances. If one has absolute freedom to design a small cost temporally connected temporal network on an underlying static network, i.e., choose the edge availabilities, then a reasonable design would be to select a rooted spanning tree and choose appropriate availabilities to construct time-respecting paths from the leaves to the root and *then* from the root back to the leaves. However, in more complicated scenarios one might not be free to *choose* edge availabilities arbitrarily but instead *specific* link availabilities might pre-exist for the network. Imagine a hostile network on a complete graph where availability of a link means a break in its security, e.g., when the guards change shifts, and only then are we able to pass a message through the link. So, if we wish to send information through the network, we may only use the times when the shifts change and it is reasonable to try and do so by using as few of these breaks as possible. In such scenarios, we may need to first verify that the pre-existing edge availabilities indeed define a temporally connected temporal network. Then, we may try to reduce the cost of the design by *removing* unnecessary (redundant) edge availabilities if possible, without loosing temporal connectivity. Consider, again, the clique network of n vertices with one time availability per edge; it is clearly temporally connected with cost $\Theta(n^2)$. However, it is not straightforward if all these edge availabilities are necessary for temporal connectivity. We resolve here the complexity of finding the maximum number of redundant labels in any given temporal graph.

1.1   The Model and Definitions

It is generally accepted to describe a network topology using a graph, the vertices and edges of which represent the communicating entities and the communication opportunities between them respectively. We consider graphs whose edge availabilities are described by sets of positive integers (labels), one set per edge.

Definition 1 (Temporal Graph). *Let $G = (V, E)$ be a (di)graph. A temporal graph on G is an ordered triplet $G(L) = (V, E, L)$, where $L = \{L_e \subseteq \mathbb{N} : e \in E\}$ is an assignment of labels[1] to the edges (arcs) of G. L is called a* labeling *of G.*

Definition 2 (Time Edge). *Let $e = \{u, v\}$ (resp. $e = (u, v)$) be an edge (resp. arc) of the underlying (di)graph of a temporal graph and consider a label $l \in L_e$. The ordered triplet (u, v, l) is called* time edge.[2]

Definition 3 (Cost of a Labeling). *Let $G(L) = (V, E, L)$ be a temporal (di)graph and L be its labeling. The* cost *of L is defined as $c(L) = \sum_{e \in E} |L_e|$.*

A basic assumption that we follow here is that when a message or an entity passes through an available link at time t, then it can pass through a subsequent link only at some time $t' > t$ and only at a time at which that link is available.

Definition 4 (Journey). *A temporal path or* journey *j from a vertex u to a vertex v ((u, v)-journey) is a sequence of time edges $(u, u_1, l_1), (u_1, u_2, l_2), \ldots, (u_{k-1}, v, l_k)$, such that $l_i < l_{i+1}$, for each $1 \leq i \leq k - 1$. We call the last time label, l_k,* arrival time *of the journey.*

Definition 5 (Foremost Journey). *A (u, v)-journey j in a temporal graph is called* foremost journey *if its arrival time is the minimum arrival time of all (u, v)-journeys' arrival times, under the labels assigned to the underlying graph's edges. We call this arrival time the* temporal distance, *$\delta(u, v)$, of v from u.*

In this work, we focus on *temporally connected* temporal graphs:

Definition 6 (Property TC). *A temporal (di)graph $G(L) = (V, E, L)$ satisfies the property TC, or equivalently L satisfies TC on G, if for any pair of vertices $u, v \in V$, $u \neq v$, there is a (u, v)-journey and a (v, u)-journey in $G(L)$. A temporal (di)graph that satisfies the property TC is called* temporally connected.

Definition 7 (Minimal Temporal Graph). *A temporal graph $G(L) = (V, E, L)$ over a (strongly) connected (di)graph is* minimal *if $G(L)$ has the property TC, and the removal of any label from any L_e, $e \in E$, results in a $G(L')$ that does not have the property TC.*

Definition 8 (Removal Profit). *Let $G(L) = (V, E, L)$ be a temporally connected temporal graph. The* removal profit *$r(G, L)$ is the largest total number of labels that can be removed from L without violating TC on G.[3]*

---

[1] The labels of an edge (arc) are the *discrete time instances* at which it is available.

[2] Note that an undirected edge $e = \{u, v\}$ is associated with $2 \cdot |L_e|$ time edges, namely both (u, v, l) and (v, u, l) for every $l \in L_e$.

[3] Here, removal of a label l from L refers to the removal of l only from a particular edge and not from all edges that are assigned label l, that is, if $l \in L_{e_1} \cap L_{e_2}$ and we remove l from both L_{e_1} and L_{e_2}, it counts as two labels removed from L.

1.2   Previous Work and Our Contribution

In recent years, there is a growing interest in distributed computing systems that are inherently dynamic. For example, temporal dynamics of network flow problems were considered in a set of pioneering papers [10,11,14,15]. The model we consider here is a direct extension of the one considered in the seminal paper of [13] and its sequel [20]. In [13], the authors consider the case of one label per edge and examines how basic graph properties change in the temporal setting. In [20], this model is extended to many labels per edge and the number of labels needed for a temporal design of a network to guarantee several graph properties with certainty is examined. The latter also defined the cost notion and, amongst other results, gave an algorithm to compute foremost journeys which can be used to decide property TC. However, the time complexity of that algorithm was *pseudo-polynomial*, as it was dominated by the *cube of the maximum label* used in the given labeling. Random edge availabilities were first considered in [1] in order to study the Expected Temporal Diameter of temporal graphs.

Here, we show that if the designer of a temporal graph can select edge availabilities freely, then an almost optimal linear-cost (in the size of the graph) design that satisfies TC can be easily obtained (cf. Sect. 3). We give an almost matching lower bound to indicate optimality. However, there are pragmatic cases where one is not free to design a temporal graph anew, but is *given* a set of possible availabilities per edge with the claim that they satisfy TC and the constraint that she may only use them or a subset of them for her design. We show that we can verify TC in *low* polynomial time (cf. Sect. 2). The *given* design may also be minimal; we partially characterize minimal designs in Sect. 4. On the other hand, there may be some labels of the initial design that can be removed without violating TC (and also result in a lower cost). In this case, how many labels can we remove at best? Our main technical result is that this problem is APX-hard, i.e. it has no PTAS unless $P = NP$. On the positive side, we show that in the case of complete graphs and random graphs, if the labels are also assigned at random, we can remove all but $O(n)$ labels.

Stochastic aspects and/or survivability of network design were also considered in [12,18,19]. An extended report of related work [3–9,16,17,21–23] can be found in our full paper (cf. Appendix).

2   Property TC Is Decidable in Low Polynomial Time

In this section, we give a simple polynomial-time algorithm which, given a temporal (di)graph $G(L)$ and a source vertex s, computes a *foremost* (s, v)-journey, for every $v \neq s$, if such a journey exists. Curiously enough, the previously known algorithm was pseudo-polynomial [20]. Our algorithm significantly improves the running time. In fact, we conjecture it is optimal.

Theorem 1. *Algorithm 1 satisfies the following, for every vertex $v \neq s$:*

(a) *If arrival_time[v] $< +\infty$, then there exists a foremost journey from s to v, the arrival time of which is exactly arrival_time[v]. This journey can be constructed by following the parent[v] pointers in reverse order.*

(b) If arrival_time[v] = +∞, then no (s, v)-journey exists.

(c) The time complexity of Algorithm 1 is dominated by the sorting time of the set of time edges (resp. time arcs).

Corollary 1. The time complexity of Algorithm 1 is $O(c(L) \cdot \log c(L))$.

Conjecture. We conjecture that any algorithm that computes journeys out of a vertex s must sort the time edges (arcs) by their labels, i.e., we conjecture that Algorithm 1 is asymptotically optimal with respect to the running time.

Note that Algorithm 1 can even compute foremost (s, v)-journeys, if they exist, that *start* from a given time $t_{start} > 0$ onward. Simply, one ignores the time edges (arcs) with labels smaller than the start time.

Algorithm 1. Foremost journey algorithm

Input: A temporal (di)graph $G(L) = (V, E, L)$ of n vertices, the set of all time edges (arcs) of which is denoted by $S(L)$; a designated source vertex $s \in V$

Output: A foremost (s, v)-journey from s to all $v \in V \setminus \{s\}$, where such a journey exists; if no (s, v)-journey exists, then the algorithm reports it.

1: Sort $S(L)$ in increasing order of labels;   // Note that $|S(L)| = c(L)$
2: Let S' be the sorted array of time edges (resp. time arcs) according to time labels;
3: $R := \{s\}$;   // The set of vertices to which s has a foremost journey
4: **for each** $v \in V \setminus \{s\}$ **do**
5:     $parent[v] := \emptyset$;
6:     $arrival\_time[v] := +\infty$;
7: Proceed sequentially in S', examining each time edge (resp. time arc) only once;
8: **for** the current time edge (resp. time arc) (a, b, l) **do**
9:     **if** $a \in R$ **and** $b \notin R$ **then**
10:        $parent[b] := a$;
11:        $arrival\_time[b] := l$;
12:        $R := R \cup \{b\}$;

---

3   Nearly Cost-Optimal Design for TC in Undirected Graphs

In this section, we study temporal design on connected undirected graphs, so that the resulting temporal graphs satisfy TC. In this scenario, the designer has absolute freedom to choose the edge availabilities of the underlying graph.

Theorem 2. (a) Given a connected undirected graph $G = (V, E)$ of n vertices, we can design a labeling L of cost $c(L) = 2(n - 1)$ that satisfies the property TC on G. L can be computed in polynomial time.

(b) For any connected undirected graph $G = (V, E)$ of $n \geq 2$ vertices and for any labeling L that satisfies the property TC on G, the cost of L is $c(L) \geq 2n - 4$.

4   Minimal Temporal Designs

Suppose now that a temporal graph on a (strongly) connected (di)graph $G = (V, E)$ is *given*[4] to a designer with the claim that it satisfies TC. If the given design is not minimal, she may wish to remove as many labels as possible, thus reducing the cost. Minimality of a design can be verified using Algorithm 1.

4.1   Minimal Designs of Non Linear Cost in the Number of Vertices

Notice that if many edges have the same label(s), we can encounter *trivial cases* of minimal temporal graphs. For example, the complete graph where all edges are assigned the same label is minimal, but there are no journeys of length larger than 1. Here, we focus on minimal temporal graphs, where minimality is not caused merely because of the use of the same labels on every edge. Consider graphs every edge of which only becomes available at most one moment in time and no two different edges become available at the same time. Are there minimal temporal graphs of the above scenario with non linear (in the size of the graph) cost? For example, any complete graph with a single label per edge, different labels to different edges, satisfies TC. Are all these $\Theta(n^2)$ labels needed for TC, i.e., are there minimal temporal complete graphs? As we prove in Theorem 4, the answer is negative. However, we give below a minimal temporal graph on n vertices with non-linear in n cost, namely with $O(n \log n)$ labels.

A minimal temporal design of $n \log n$ cost

Definition 9 (Hypercube Graph). *The k-hypercube graph, commonly denoted Q_k, is the k-regular graph of 2^k vertices and $2^{k-1} \cdot k$ edges.*

Theorem 3. *There exists an infinite class of minimal temporal graphs on n vertices with $\Theta(n \cdot \log n)$ edges and $\Theta(n \cdot \log n)$ labels, such that different edges have different labels.*

Sketch of Proof. We present a minimal temporal graph on the hypercube. Consider Protocol 2 for labeling the edges of $G = Q_k$. The temporal graph, $G(L)$, that this labeling procedure produces on the hypercube is minimal.   □

Cliques of at Least 4 Vertices are not Minimal. The complete graph on n vertices, K_n, with a labeling L that assigns a single, different for every edge, label per edge is an interesting case, since $K_n(L)$ always satisfies TC. However, it is not minimal as the theorem below shows.

Theorem 4. *Let $n \in \mathbb{N}$, $n \geq 4$ and denote by K_n the complete graph on n vertices. Any labeling L that assigns a single label to every edge of K_n, different label per edge produces a temporal graph $K_n(L)$ that is not minimal. In fact, $\exists S \subseteq \cup_{e \in E(K_n)} L_e$, $|S| = \lfloor \frac{n}{4} \rfloor$, such that when we remove all the labels of S*

---

[4] In this scenario, the designer is allowed to only use the given set of edge availabilities, or a subset of them.

Protocol 2. Labeling the hypercube graph, $G = Q_k$

Consider the k dimensions of the hypercube $G = Q_k$, $x_1, x_2, \ldots, x_k$;
for $i = 1 \ldots k$ **do**
    Let $X_i := \{e_{i1}, e_{i2}, \ldots, e_{i2^{k-1}}\}$ be the list of edges in dimension x_i, in an arbitrary order;
    Let L_i be the (sorted from smallest to largest) list of labels $L_i := \{(i-1) \cdot 2^{k-1} + 1, (i-1) \cdot 2^{k-1} + 2, \ldots, i \cdot 2^{k-1}\}$;
for $i = 1 \ldots k$ **do**
    for $j = 1 \ldots 2^{k-1}$ **do**
        Assign the (current) first label of L_i to the (current) first edge of X_i ;
        Remove the (current) first label of L_i from the list;
        Remove the (current) first edge of X_i from the list;
return the produced temporal graph, $G(L)$;

---

from L, the resulting temporal graph still satisfies TC. Note that by the union $\cup_{e \in E(K_n)} L_e$ we denote the multiset *of all labels used in L.*

Proof. The proof is divided in two parts, as follows:

(a) We show that the theorem holds for K_4. Without loss of generality, we use labels 1 to 6, one label per edge, and show that we can always remove a label and still satisfy TC. The proof requires an exhaustive check of 720 permutations of the labels and is done via a *computer program* (code can be found online here: http://cgi.csc.liv.ac.uk/~akridel/research-results.html).

(b) Now, consider the complete graph on $n \geq 4$ vertices, $K_n = (V, E)$. Partition V arbitrarily into $\lceil \frac{n}{4} \rceil$ subsets $V_1, V_2, \ldots, V_{\lceil \frac{n}{4} \rceil}$, such that $|V_i| = 4, \forall i = 1, 2, \ldots, \lceil \frac{n}{4} \rceil - 1$ and $|V_{\lceil \frac{n}{4} \rceil}| \leq 4$. In each 4-clique defined by V_i, $i = 1, 2, \ldots, \lfloor \frac{n}{4} \rfloor$, we can remove a "redundant" label, as shown in (a). The resulting temporal graph on K_n still preserves TC since for every ordered pair of vertices $u, v \in V$:
   – if u, v are in the same V_i, $i = 1, 2, \ldots, \lfloor \frac{n}{4} \rfloor$, then there is a (u, v)-journey that uses time edges within the 4-clique on V_i, as proven in (a).
   – if $u \in V_i$ and $v \in V_j$, $i \neq j$, then there is a (u, v)-journey that uses the (direct) time edge on $\{u, v\}$.   □

4.2 Computing the Removal Profit is APX-hard

Recall that the removal profit is the largest number of labels that can be removed from a temporally connected graph without destroying TC. We now show that it is hard to arbitrarily approximate the value of the removal profit for an arbitrary graph, i.e., there exists no PTAS[5] for this problem, unless P=NP. It is worth noting here that, in our hardness proof below, we consider undirected graphs.

    We prove our hardness result by providing an approximation preserving polynomial reduction from a variant of the maximum satisfiability problem, namely

---

[5] PTAS stands for Polynomial-Time Approximation Scheme.

from the *monotone Max-XOR(3)* problem. Consider a monotone XOR-boolean formula ϕ with variables $x_1, x_2, \ldots, x_n$.[6] The clause $\alpha = (x_i \oplus x_j)$ is XOR-satisfied by a truth assignment τ iff $x_i \neq x_j$ in τ. The number of clauses of ϕ that are XOR-satisfied in τ is denoted by $|\tau(\phi)|$. If every variable x_i appears in exactly k XOR-clauses in ϕ, then ϕ is called a *monotone XOR(k)* formula. The *monotone Max-XOR(k)* problem is, given a monotone XOR(k) formula ϕ, to compute a truth assignment τ of the variables $x_1, x_2, \ldots, x_n$ that XOR-satisfies the largest possible number of clauses, i.e., an assignment τ such that $|\tau(\phi)|$ is maximized. The monotone Max-XOR(3) problem essentially encodes the *Max-Cut* problem on 3-regular (cubic) graphs, which is known to be APX-hard [2].

Lemma 1 [2]. *The monotone Max-XOR(3) problem is APX-hard.*

Now we provide our reduction from the monotone Max-XOR(3) problem to the problem of computing $r(G, L)$. Let ϕ be an arbitrary instance of monotone Max-XOR(3) with n variables $x_1, x_2, \ldots, x_n$ and m clauses. Since every variable x_i appears in ϕ in exactly 3 clauses, it follows that $m = \frac{3}{2}n$. We will construct from ϕ a graph $G_\phi = (V_\phi, E_\phi)$ and a labeling L_ϕ of G_ϕ.

For every $i = 1, 2, \ldots, n$ we construct the graph $G_{\phi,i}$ and the labeling $L_{\phi,i}$ of Fig. 1. In this figure, the labels of every edge in $L_{\phi,i}$ are drawn next to the edge. We call the induced subgraph of $G_{\phi,i}$ on the 4 vertices $\{s^{x_i}, u_0^{x_i}, w_0^{x_i}, v_0^{x_i}\}$ the *base* of $G_{\phi,i}$. Also, for every $p \in \{1, 2, 3\}$, we call the induced subgraph of $G_{\phi,i}$ on the 4 vertices $\{t_p^{x_i}, u_p^{x_i}, w_p^{x_i}, v_p^{x_i}\}$ the *pth branch* of $G_{\phi,i}$. Finally, we call the edges $\{u_0^{x_i}, w_0^{x_i}\}$ and $\{w_0^{x_i}, v_0^{x_i}\}$ the *transition edges* of the base of $G_{\phi,i}$ and, for every $p \in \{1, 2, 3\}$, we call the edges $\{u_p^{x_i}, w_p^{x_i}\}$ and $\{w_p^{x_i}, v_p^{x_i}\}$ the *transition edges* of the *pth branch* of $G_{\phi,i}$. For every $p \in \{1, 2, 3\}$ we associate the *pth* appearance of the variable x_i in a clause of ϕ with the *pth branch* of $G_{\phi,i}$.

We continue the construction of $G_{\phi,i}$ and $L_{\phi,i}$ as follows. First, we add an edge between any possible pair of vertices $w_p^{x_i}, w_q^{x_j}$, where $p, q \in \{0, 1, 2, 3\}$ and $i, j \in \{1, 2, \ldots, n\}$, and we assign to this new edge $e = \{w_p^{x_i}, w_q^{x_j}\}$ the unique

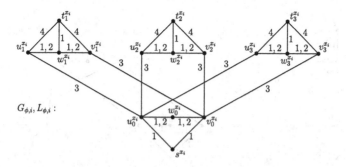

Fig. 1. The gadget $G_{\phi,i}$ for the variable x_i.

---

[6] A monotone XOR-boolean formula is a conjunction of XOR-clauses of the form $(x_i \oplus x_j)$, where no variable is negated.

label $L_\phi(e) = \{7\}$. Note here that we add this edge $\{w_p^{x_i}, w_q^{x_j}\}$ also in the case where $i = j$ (and $p \neq q$). Moreover, we add an edge between any possible pair of vertices $t_p^{x_i}, t_q^{x_j}$, where $i \neq j$, $i,j \in \{1,2,\ldots,n\}$, and $p,q \in \{1,2,3\}$. We assign to this new edge $e = \{t_p^{x_i}, t_q^{x_j}\}$ the unique label $L_\phi(e) = \{7\}$.

Furthermore we add a new vertex t_0 which is adjacent to vertex $w_0^{x_n}$ and to all vertices in the set $\{s^{x_i}, t_1^{x_i}, t_2^{x_i}, t_3^{x_i}, u_p^{x_i}, v_p^{x_i} : 1 \leq i \leq n, \ 0 \leq p \leq 3\}$. First we assign to the edge $\{t_0, w_0^{x_n}\}$ the unique label $L_\phi(\{t_0, w_0^{x_n}\}) = \{5\}$. Furthermore, for every vertex $t_p^{x_i}$, where $1 \leq i \leq n$ and $1 \leq p \leq 3$, we assign to the edge $\{t_0, t_p^{x_i}\}$ the unique label $L_\phi(\{t_0, t_p^{x_i}\}) = \{5\}$. Finally, for each of the vertices $z \in \{s^{x_i}, u_p^{x_i}, v_p^{x_i} : 1 \leq i \leq n, \ 0 \leq p \leq 3\}$ we assign to the edge $\{t_0, z\}$ the unique label $L_\phi(\{t_0, z\}) = \{6\}$. The addition of the vertex t_0 and the labels of the (dashed) edges incident to t_0 are illustrated in Figure 2(a).

Consider now a clause $\alpha = (x_i \oplus x_j)$ of ϕ. Assume that the variable x_i (resp. x_j) of the clause α corresponds to the pth (resp. qth) appearance of x_i (resp. x_j) in ϕ. Then we identify the vertices $u_p^{x_i}, v_p^{x_i}, w_p^{x_i}, t_p^{x_i}$ of the pth branch of $G_{\phi,i}$ with the vertices $v_q^{x_i}, u_q^{x_i}, w_q^{x_i}, t_q^{x_i}$ of the qth branch of $G_{\phi,j}$, respectively (cf. Figure 2(b)). This completes the construction of G_ϕ and its labeling L_ϕ.

Denote the vertex sets $A = \{s^{x_i}, u_p^{x_i}, v_p^{x_i} : 1 \leq i \leq n, \ 0 \leq p \leq 3\}$, $B = \{w_p^{x_i} : 1 \leq i \leq n, \ 0 \leq p \leq 3\}$, and $C = \{t_p^{x_i} : 1 \leq i \leq n, \ 1 \leq p \leq 3\}$. Note that $V_\phi = A \cup B \cup C \cup \{t_0\}$. Furthermore, for every $i \in \{1,2,\ldots,n\}$ and every $p \in \{1,2,3\}$ we define for simplicity of notation the temporal paths $P_{i,p} = (s^{x_i}, u_0^{x_i}, u_p^{x_i}, t_p^{x_i})$ and $Q_{i,p} = (s^{x_i}, v_0^{x_i}, v_p^{x_i}, t_p^{x_i})$. For every $i \in \{1,2,\ldots,n\}$ the graph $G_{\phi,i}$ has 16 vertices. Furthermore, for every $p \in \{1,2,3\}$, the 4 vertices of the pth branch of $G_{\phi,i}$ also belong to a branch of $G_{\phi,j}$, for some $j \neq i$. Therefore, together with the vertex t_0, the graph G_ϕ has in total $10n+1$ vertices.

To provide some intuition about the correctness of Theorem 5, we now briefly describe how we can construct a labeling L of G_ϕ by removing $9n+k$ labels from L_ϕ, given a truth assignment τ of ϕ with $|\tau(\phi)| \geq k$. First we keep in L all labels of L_ϕ on the edges incident to t_0. Furthermore we keep in L the label $\{7\}$ of all the edges $\{t_p^{x_i}, t_q^{x_j}\}$ and the label $\{7\}$ of all the edges $w_p^{x_i} w_q^{x_j}$. Moreover we keep

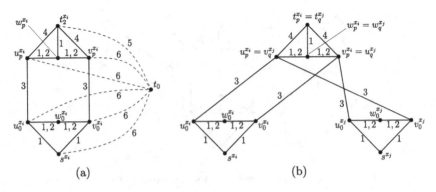

(a)   (b)

Fig. 2. (a) The addition of vertex t_0. There exists in G_ϕ also the edge $\{t_0, w_0^{x_n}\}$ with label $L_\phi(\{t_0, w_0^{x_n}\}) = \{5\}$. (b) The gadget for the clause $(x_i \oplus x_j)$.

in L the label $\{1\}$ of all the edges $\{t_p^{x_i}, w_p^{x_i}\}$. Let now $i = 1, 2, \ldots, n$. If $x_i = 0$ in τ, we keep in L the labels of the edges of the paths $P_{i,1}, P_{i,2}, P_{i,3}$, as well as the label 1 of the edge $\{v_0^{x_i}, w_0^{x_i}\}$ and the label 2 of the edge $\{w_0^{x_i}, u_0^{x_i}\}$. Otherwise, if $x_i = 1$ in τ, we keep in L the labels of the edges of the paths $Q_{i,1}, Q_{i,2}, Q_{i,3}$, as well as the label 1 of the edge $\{u_0^{x_i}, w_0^{x_i}\}$ and the label 2 of the edge $\{w_0^{x_i}, v_0^{x_i}\}$.

We now continue the labeling L as follows. Consider an arbitrary clause $\alpha = (x_i \oplus x_j)$ of ϕ. Assume that the variable x_i (resp. x_j) of α corresponds to the pth (resp. to the qth) appearance of variable x_i (resp. x_j) in ϕ. Then, by the construction of G_ϕ, the pth branch of $G_{\phi,i}$ coincides with the qth branch of $G_{\phi,j}$, i.e., $u_p^{x_i} = v_q^{x_j}$, $v_p^{x_i} = u_q^{x_j}$, $w_p^{x_i} = w_q^{x_j}$, and $t_p^{x_i} = t_q^{x_j}$. Let α be XOR-satisfied in τ, i.e., $x_i = \overline{x_j}$. If $x_i = \overline{x_j} = 0$ (i.e., $x_i = 0$ and $x_j = 1$) then we keep in L the label 1 of the edge $\{v_p^{x_i}, w_p^{x_i}\}$ and the label 2 of the edge $\{w_p^{x_i}, u_p^{x_i}\}$. In the symmetric case, where $x_i = \overline{x_j} = 1$ (i.e., $x_i = 1$ and $x_j = 0$), we keep in L the label 1 of the edge $\{u_p^{x_i}, w_p^{x_i}\}$ and the label 2 of the edge $\{w_p^{x_i}, v_p^{x_i}\}$. Let now α be XOR-unsatisfied in τ, i.e., $x_i = x_j$. Then, in both cases where $x_i = x_j = 0$ and $x_i = x_j = 1$, we keep in L the label 1 of both edges $\{v_p^{x_i}, w_p^{x_i}\}$ and $\{w_p^{x_i}, u_p^{x_i}\}$. This finalizes the construction of L from the truth assignment τ of ϕ.

Theorem 5. *There is an assignment τ of ϕ with $|\tau(\phi)| \geq k$ iff $r(G, L) \geq 9n + k$.*

Theorem 6. *The problem of computing $r(G, L)$ on a graph G is APX-hard.*

Proof. Denote now by $\mathrm{OPT}_{\text{mon-Max-XOR(3)}}(\phi)$ the greatest number of clauses that can be simultaneously XOR-satisfied by a truth assignment of ϕ. Then Theorem 5 implies that $r(G_\phi, L_\phi) \geq 9n + \mathrm{OPT}_{\text{mon-Max-XOR(3)}}(\phi)$. Note that a random truth assignment XOR-satisfies each clause of ϕ with probability $\frac{1}{2}$, and thus there exists an assignment τ that XOR-satisfies at least $\frac{m}{2}$ clauses of ϕ. Therefore $\mathrm{OPT}_{\text{mon-Max-XOR(3)}}(\phi) \geq \frac{m}{2} = \frac{3}{4}n$, and thus, $n \leq \frac{4}{3}\mathrm{OPT}_{\text{mon-Max-XOR(3)}}(\phi)$. Assume that there is a PTAS for computing $r(G, L)$. Then, for every $\varepsilon > 0$ we can compute in polynomial time a labeling $L \subseteq L_\phi$ for the graph G_ϕ, such that $|L_\phi \setminus L| \geq (1 - \varepsilon) \cdot r(G_\phi, L_\phi)$. Given such a labeling $L \subseteq L_\phi$ we can compute by the sufficiency part ($\Leftarrow$) of the proof of Theorem 5 a truth assignment τ of ϕ so that $|L_\phi \setminus L| \leq 9n + |\tau(\phi)|$, i.e., $|\tau(\phi)| \geq |L_\phi \setminus L| - 9n$. Therefore it follows that:

$$|\tau(\phi)| \geq (1 - \varepsilon) \cdot r(G_\phi, L_\phi) - 9n$$
$$\geq (1 - \varepsilon) \cdot \left(9n + \mathrm{OPT}_{\text{mon-Max-XOR(3)}}(\phi)\right) - 9n$$
$$\geq (1 - \varepsilon) \cdot \left(\mathrm{OPT}_{\text{mon-Max-XOR(3)}}(\phi)\right) - 9\varepsilon \cdot \frac{4}{3}\mathrm{OPT}_{\text{mon-Max-XOR(3)}}(\phi)$$
$$= (1 - 13\varepsilon) \cdot \left(\mathrm{OPT}_{\text{mon-Max-XOR(3)}}(\phi)\right)$$

That is, assuming a PTAS for computing $r(G, L)$, we obtain a PTAS for the monotone Max-XOR(3) problem, which is a contradiction by Lemma 1, unless $P = NP$. So, computing $r(G, L)$ on an undirected graph G is APX-hard. $\square$

4.3 Random Labelings on Dense Graphs Have High Removal Profit

We show here that dense graphs with random labels satisfy TC and have a very high removal profit with high probability (whp).

Definition 10. *We call* normalized uniform random temporal graph *any graph on* $n \in \mathbb{N}$ *vertices, each edge of which receives exactly one label uniformly at random, and independently of other edges, from the set* $\{1, \ldots, n\}$.

Theorem 7. *(a) In the normalized uniform random temporal clique of n vertices, we can delete all but $2n + O(\log n)$ labels without violating TC whp.*

(b) Let $G = (V, E)$ be an instance of the Erdös-Renyi model of random graphs, $G_{n,p}$, with $p \geq \frac{a\sqrt{n}\log n}{n}$, where a is constant, and consider a normalized uniform random temporal graph, $G(L)$, on G. We can delete all but $2n + O(\sqrt{n})$ labels of $G(L)$ without violating TC whp.

Sketch of Proof. We provide a sketch of the proof of (a). Partition the set of available labels $\{1, 2, \ldots, n\}$ into 4 consecutive equisized subsets $A_1, \ldots, A_4$. Each edge receives a single random label l, with $Pr[l \in A_i] = \frac{1}{4}$, $\forall i = 1, 2, 3, 4$. Color *green(g)*, *yellow(y)*, *blue(b)* and *red(r)* the edges that are assigned a label in A_1, A_2, A_3 and A_4 respectively. A *temporal router* is a subgraph of the clique consisting of a central vertex with a number of yellow incident edges and a number of blue incident edges. Fix a vertex u of the graph. By use of Chernoff bounds, we show the following:

Lemma 2. *There is a set S_1 of at least $\frac{n}{4}$ yellow edges incident to u and a set S_2 of at least $\frac{n}{4}$ blue edges incident to u, with probability at least $1 - 2e^{-\frac{n}{16}}$.*

Conditioning on the above property of u, we arbitrarily select a subset D_i of S_i with $|D_i| = \alpha \log n$, $i = 1, 2$. $R = D_1 \cup D_2 \cup \{u\}$ is then a $O(\log n)$-size temporal router.

Lemma 3. *Any vertex $w \in V \setminus R$ has an incident g edge to a vertex in D_1 and an incident r edge to a vertex in D_2 with probability at least $1 - 2e^{-\frac{\alpha \log n}{4}}$.*

Using Lemma 3, we show that whp, we can remove all the labels from the random labeling on the graph except for the labels on the edges of the "router" and the two incident edges of any $w \in V$, one g connecting it to a vertex in D_1 and one r connecting it to a vertex in D_2, and still satisfy the property TC. $\qquad \square$

Acknowledgments. We wish to thank Thomas Gorry for co-implementing the code used in the proof of Theorem 4.

References

1. Akrida, E.C., Gąsieniec, L., Mertzios, G.B., Spirakis, P.G.: Ephemeral networks with random availability of links: Diameter and connectivity. In: Proceedings of the 26th ACM Symposium on Parallelism in Algorithms and Architectures (SPAA) (2014)
2. Alimonti, P., Kann, V.: Hardness of approximating problems on cubic graphs. In: Bongiovanni, G., Bovet, D.P., Di Battista, G. (eds.) CIAC 1997. LNCS, vol. 1203, pp. 288–298. Springer, Heidelberg (1997)

3. Avin, C., Koucký, M., Lotker, Z.: How to explore a fast-changing world (cover time of a simple random walk on evolving graphs). In: Aceto, L., Damgård, I., Goldberg, L.A., Halldórsson, M.M., Ingólfsdóttir, A., Walukiewicz, I. (eds.) ICALP 2008, Part I. LNCS, vol. 5125, pp. 121–132. Springer, Heidelberg (2008)
4. Bui-Xuan, B.-M., Ferreira, A., Jarry, A.: Computing shortest, fastest, and foremost journeys in dynamic networks. Int. J. Found. Comput. Sci. 14(2), 267–285 (2003)
5. Casteigts, A., Flocchini, P.: Deterministic Algorithms in Dynamic Networks: Formal Models and Metrics. Defence R&D Canada, Technical report, April 2013
6. Casteigts, A., Flocchini, P.: Deterministic Algorithms in Dynamic Networks: Problems, Analysis, and Algorithmic Tools. Defence R&D Canada, Technical report, April 2013
7. Casteigts, A., Flocchini, P., Quattrociocchi, W., Santoro, N.: Time-varying graphs and dynamic networks. Int. J. Parallel Emergent Distrib. Syst. (IJPEDS) 27(5), 387–408 (2012)
8. Clementi, A.E.F., Macci, C., Monti, A., Pasquale, F., Silvestri, R.: Flooding time of edge-markovian evolving graphs. SIAM J. Discrete Math. (SIDMA) 24(4), 1694–1712 (2010)
9. Dutta, C., Pandurangan, G., Rajaraman, R., Sun, Z., Viola, E.: On the complexity of information spreading in dynamic networks. In: Proceedings of the 24th Annual ACM-SIAM Symposium on Discrete Algorithms (SODA), pp. 717–736 (2013)
10. Fleischer, L., Skutella, M.: Quickest flows over time. SIAM J. Comput. 36(6), 1600–1630 (2007)
11. Fleischer, L., Tardos, É.: Efficient continuous-time dynamic network flow algorithms. Oper. Res. Lett. 23(3–5), 71–80 (1998)
12. Gupta, A., Krishnaswamy, R., Ravi, R.: Online and stochastic survivable network design. SIAM J. Comput. 41(6), 1649–1672 (2012)
13. Kempe, D., Kleinberg, J.M., Kumar, A.: Connectivity and inference problems for temporal networks. In: Proceedings of the 32nd Annual ACM symposium on Theory of computing (STOC), pp. 504–513 (2000)
14. Klinz, B., Woeginger, G.J.: One, two, three, many, or: complexity aspects of dynamic network flows with dedicated arcs. Oper. Res. Lett. 22(4–5), 119–127 (1998)
15. Koch, R., Nasrabadi, E., Skutella, M.: Continuous and discrete flows over time - A general model based on measure theory. Math. Meth. of OR 73(3), 301–337 (2011)
16. Kontogiannis, S., Zaroliagis, C.: Distance oracles for time-dependent networks. In: Esparza, J., Fraigniaud, P., Husfeldt, T., Koutsoupias, E. (eds.) ICALP 2014. LNCS, vol. 8572, pp. 713–725. Springer, Heidelberg (2014)
17. Kuhn, F., Lynch, N.A., Oshman, R.: Distributed computation in dynamic networks. In: Proceedings of the 42nd Annual ACM Symposium on Theory of Computing (STOC), pp. 513–522 (2010)
18. Lau, L.C., Naor, J., Salavatipour, M.R., Singh, M.: Survivable network design with degree or order constraints. SIAM J. Comput. 39(3), 1062–1087 (2009)
19. Lau, L.C., Singh, M.: Additive approximation for bounded degree survivable network design. SIAM J. Comput. 42(6), 2217–2242 (2013)
20. Mertzios, G.B., Michail, O., Chatzigiannakis, I., Spirakis, P.G.: Temporal network optimization subject to connectivity constraints. In: Fomin, F.V., Freivalds, R.U., Kwiatkowska, M., Peleg, D. (eds.) ICALP 2013, Part II. LNCS, vol. 7966, pp. 657–668. Springer, Heidelberg (2013)

21. Michail, O., Chatzigiannakis, I., Spirakis, P.G.: Causality, influence, and computation in possibly disconnected synchronous dynamic networks. In: Baldoni, R., Flocchini, P., Binoy, R. (eds.) OPODIS 2012. LNCS, vol. 7702, pp. 269–283. Springer, Heidelberg (2012)
22. O'Dell, R., Wattenhofer, R.: Information dissemination in highly dynamic graphs. In: Proceedings of the 2005 Joint Workshop on Foundations of Mobile Computing (DIALM-POMC), pp. 104–110 (2005)
23. Scheideler, C.: Models and techniques for communication in dynamic networks. In: Alt, H., Ferreira, A. (eds.) STACS 2002. LNCS, vol. 2285, pp. 27–49. Springer, Heidelberg (2002)

Efficient Vertex-Label Distance Oracles
for Planar Graphs

Shay Mozes$^{(\boxtimes)}$ and Eyal E. Skop

Efi Arazi School of Computer Science, The Interdisciplinary Center Herzliya,
Herzliya, Israel
smozes@idc.ac.il, shaymozes@gmail.com, eyalskop@gmail.com

Abstract. We consider distance queries in vertex labeled planar graphs. For any fixed $0 < \epsilon \leq 1/2$ we show how to preprocess an undirected planar graph with vertex labels and edge lengths to answer queries of the following form. Given a vertex u and a label λ return a $(1 + \epsilon)$-approximation of the distance between u and its closest vertex with label λ. The query time of our data structure is $O(\lg \lg n + \epsilon^{-1})$, where n is the number of vertices. The space and preprocessing time of our data structure are nearly linear. We give a similar data structure for directed planar graphs with slightly worse performance. The best prior result for the undirected case has similar space and preprocessing bounds, but exponentially slower query time. No nontrivial results were previously considered for the directed case.

1 Introduction

Imagine you are driving your car and suddenly notice you are about to run out of gas. What should you do? Obviously, you should find the closest gas station. This is the *vertex-to-label distance query problem*. Various software applications like Waze and Google maps attempt to provide such a functionality. The idea is to preprocess the locations of service providers, such as gas stations, hospitals, pubs and metro stations in advance, so that when a user, whose location is not known a priori, asks for the distance to the closest service provider, the information can be retrieved as quickly as possible.

We study this problem from a theoretical point of view. We model the network as a planar graph with labeled vertices (some papers refer to labels as colors). For example, a vertex can be labeled as a gas station. We study distance oracles for such graphs. A *vertex-label distance oracle* is a data structure that represents the input graph and can be queried for the distance between any vertex and the closest vertex with a desired label. We consider approximate distance oracles, which, for any given fixed parameter $\epsilon > 0$, return a distance

This work was partially supported by Israel Science Foundation grant 794/13 and by the Israeli ministry of absorption.

A full version of this paper, including figures, can be found in http://arxiv.org/abs/1504.04690.

L. Sanità and M. Skutella (Eds.): WAOA 2015, LNCS 9499, pp. 97–109, 2015.
DOI: 10.1007/978-3-319-28684-6_9

estimate that is at least the true distance queried, and at most $(1 + \epsilon)$ times the true distance (this is also called $(1 + \epsilon)$-*stretch*). One would like an oracle with the following properties; queries should be answered quickly, the oracle should consume little space, and the construction of the oracle should take as little time as possible. We use the notation $\langle O(S(n))_{space}, O(T(n))_{time} \rangle$ to express the space requirement and query time of a distance oracle.

Our Results and Approach. Our results are summarized as follows:

Theorem 1. *A $(1 + \epsilon)$-stretch $\langle O(\epsilon^{-1} n \lg n)_{space}, O(\lg \lg n + \epsilon^{-1})_{time} \rangle$ vertex-label distance oracle can be constructed in $O(\epsilon^{-2} n \lg^3 n)$ time for an undirected planar graph with n vertices.*

Theorem 2. *A $(1 + \epsilon)$-stretch $\langle O(\epsilon^{-1} n \lg n \lg(nN))_{space}, O(\lg \lg n \lg \lg (nN) + \epsilon^{-1})_{time} \rangle$ vertex-label distance oracle can be constructed in $O(\epsilon^{-2} n \lg^3 n \lg(nN))$ time for a directed planar graph with n vertices and maximum integer arc length N.*

Consider a vertex-to-vertex distance oracle and a graph with label set L. If the oracle works for general directed graphs then the vertex-to-label problem can be solved easily; add a distinct apex v_λ for each label $\lambda \in L$, and connect every λ-labeled vertex to v_λ with a zero length arc. Finding the distance from a vertex u to label λ is now equivalent to finding the distance between u and v_λ. This approach presents two main difficulties when designing efficient oracles for planar graphs. First, adding apices breaks planarity. In particular, it affects the separability of the graph. Thus, the reduction does not work with oracles that depend on planarity or on the existence of separators, which are more efficient than oracles for general graphs. Second, the reduction uses *directed* arcs, so it is unsuitable for oracles for undirected graphs. Using arcs in the reduction is crucial since connecting an apex with undirected zero length edges changes the distances in the graph. This is because the apex can be used to teleport between vertices with the same label.

Our contribution is in realizing and showing that the internal workings of vertex-to-vertex distances oracles for planar graphs due to Thorup [11] can be extended to support vertex labels. Achieving this modification is non-trivial since introducing the apices needs to be done in a manner that guarantees correctness without compromising efficiency. Thorup's oracles rely on the existence of fundamental cycle separators in planar graphs, a property that breaks when apices are added to the graph. We observe, however, that once the graph is separated, Thorup's oracle does not depend on planarity. We therefore postpone the addition of the apices till a later stage in the construction of the distance oracle, when the graph has already been separated. We show that, nonetheless, approximate distances from any vertex to any label in the entire graph can be efficiently approximated. Furthermore, we extend a technique of Thorup, originally intended to reduce the amount of information stored for a single vertex, to handle all vertices with the same label as if they were a single vertex.

Related Work. We first summarize related work on approximate vertex-vertex distance oracles for planar graphs. Thorup [11] gave a $\langle O(\epsilon^{-1}n \lg n \lg(nN))_{space},$ $O(\lg \lg(nN) + \epsilon^{-1})_{time}\rangle$ stretch $(1 + \epsilon)$ directed distance oracle, and a $\langle O(\epsilon^{-1}n \lg n)_{space}, O(\epsilon^{-1})_{time}\rangle$ undirected (simplified) distance oracle. Our result is based on Thorup's oracles, which are described in Sect. 3. Klein [7] independently gave an undirected distance oracle with the same bounds. Kawarabayashi, Klein and Sommer [5] have shown a $\langle O(n)_{space},$ $O(\epsilon^{-2}\lg^{-2}(n))_{time}\rangle$ undirected $(1 + \epsilon)$-stretch distance oracle constructed in $O(n \lg^2 n)$ time, inspired by [11]. For a parameter r they give a trade-off of $\langle O(\frac{\epsilon^{-1}n \lg n}{\sqrt{r}})_{space}, O(r + \sqrt{r}\epsilon^{-1} \lg n)_{time}\rangle$ oracle algorithms. Kawarabayashi, Sommer and Thorup [6] have shown better tradeoffs for undirected oracles. For the case where $N \in poly(n)$, they achieve $\langle O^*(n \lg n)_{space}, O^*(\epsilon^{-1})_{time}\rangle$ oracle, where O^* hides $\lg(\epsilon^{-1})$ and $\lg^*(n)$ factors.

The vertex-to-label distance query problem was introduced by Hermelin, Levy, Weimann and Yuster [4]. For any integer $k \geq 2$, they gave a $(4k - 5)$-stretch $\langle O(kn^{1+1/k})_{space}, O(k)_{time}\rangle$ vertex-label distance oracle (expected space) for undirected general (i.e., non-planar) graphs. This is not efficient when the number l of distinct labels is $o(n^{1/k})$. They also presented a $(2^k - 1)$-stretch $\langle O(knl^{1/k})_{space}, O(k)_{time}\rangle$ undirected oracle, and showed how to maintain label changes in sub-linear time. Chechik [2] improved the latter two results to $(4k-5)$-stretch and similar space/time bounds.

For planar graphs, Li, Ma and Ning [8], building on [7], construct a $(1 + \epsilon)$-stretch vertex-labeled oracle with $\langle O(\epsilon^{-1}n \lg n)_{space}, O(\epsilon^{-1} \lg n \lg \rho)_{time}\rangle$ bounds for undirected graphs. Here, ρ is the radius of the graph, which can be $\theta(n)$. It is also shown in [8] how to avoid the $\lg \rho$ factor when $\rho = O(\lg n)$. The construction time of their oracle is $O(\epsilon^{-1}n \log^2 n)$. In comparison, the query time of our undirected oracle is lower by roughly a $\log^2 n$ factor. I.e., exponentially faster. The space requirement is the same, but our preprocessing is slower by an $\epsilon^{-1} \log n$ factor.

It was recently brought to our attention that Łącki, Oćwieja, Pilipczuk, Sankowski, and Zych [10] developed dynamic vertex-labeled distance oracles for undirected general and planar graphs, and used them to maintain approximate solutions for dynamic Steiner and subgraph TSP problems. They describe a generic scheme for converting certain undirected distance oracles into dynamic undirected vertex-label distance oracle. Applying their scheme to one of the slower variants of Thorup's distance oracles, they obtain a $\langle O(\epsilon^{-1}n \log n \log(nN))_{space}, O(\epsilon^{-1} \log n \log(nN))_{time}\rangle$ $(1+\epsilon)$-stretch undirected vertex-labeled distance oracle that also supports merging labels. Our result for undirected graphs has exponentially faster query time, but our preprocessing is slower by a constant ϵ^{-1} factor. Our work does not address dynamic operations on the labels. We believe, however, that extending our scheme to support merges in logarithmic amortized time is possible using similar arguments to [10].

Another recent related work is the one by Abraham, Chechik, Krauthgamer and Wieder [1], who considered approximate nearest neighbor search in planar graph metrics. This is the special case of vertex-labeled distance oracle with only one label. for this weaker problem they obtain a data structure whose size is nearly linear in the number of labeled vertices. However, they assume an *exact* vertex-to-vertex distance oracle is provided.

To the best of our knowledge, no non-trivial directed vertex-label distance oracles were proposed prior to the current work.

Roadmap. In this extended abstract we focus on the undirected case. The remainder of this paper is organized as follows. We describe the scheme of the vertex-to-vertex distance oracle of Thorup in Sect. 3. In Sect. 4, we describe a vertex-labeled oracle for undirected planar graphs. Our description goes into some of the details that are missing in the treatment of the undirected case in [11]. See the full version for ellaboration. Due to space constraints, our vertex-labeled distance oracle for directed planar graphs is described in the full version of the paper. Its construction is similar to the undirected oracle, but relies on some additional reductions from [11] that we use without change.

2 Preliminaries

Let $V(G)$ denote the vertex set of a graph G. We use the terms arcs and edges to distinguish directed and undirected graphs. Let $A(G)$ ($E(G)$) denote the arc (edge) set of a directed (undirected) graph. We denote the concatenation of two paths P_1 and P_2 that share an endpoint by $P_1 \circ P_2$.

For a simple path Q and a vertex set $U \subseteq V(Q)$ with $|U| \geq 2$, we define $\bar{Q}$, the *reduction* of Q to U as follows. Repeatedly apply the following procedure to Q. Let wv be an edge of Q s.t. $v \notin U$. Contract wv, and add the length of wv to the length of the other edge of Q incident to w. Note that $|V(\bar{Q})| = O(|U|)$.

Let T be a rooted spanning tree of a graph G. For $u \in V(G)$, let $T[u]$ denote the unique root-to-u path in T. The *fundamental cycle* of $e = (u_1, u_2) \notin E(T)$ is the (not necessarily simple) undirected cycle composed of $E(T[u_1]), E(T[u_2])$, and e.

Let $L = \{\lambda_i\}_{i=1}^l$ be a set of l labels. A vertex-labeled graph is a graph $G = (V, A)$, equipped with a function $f : V \to L$. Let $V_\lambda = \{v \in V(G) | f(v) = \lambda\}$ to be set of vertices with label λ.

Let G be a graph with arc lengths. For $u, v \in V(G)$, let $\delta_G(u, v)$ denote the u-to-v distance in G. For a vertex-labeled G, we define $\delta_G(u, \lambda) = \min_{w \in V_\lambda} \delta_G(u, w)$.

We assume basic familiarity with planar graphs. In particular, it is well known that if G is planar then $|A(G)| = O(|V(G)|)$, and that a simple cycle separates a planar graph G into an interior and an exterior parts.

A *vertex-label distance oracle* is a data structure that, given a vertex $v \in V$ and a label $\lambda \in L$, outputs an (approximation of) $\delta_G(v, \lambda)$. We note that this problem is a generalization of the basic distance oracle problem in which each vertex is given a unique label. Constructing an $O(nl)$-space vertex-label distance oracle is trivial. Simply precompute and store the distance between each vertex and each possible label. The goal is, therefore, to devise an oracle which requires substantially less than nl space, while allowing for fast queries.

3   Thorup's Approximate Distance Oracle

In this section we outline the distance oracle of Thorup [11]. This is necessary for understanding our results. The oracle we describe differs from the original in [11] in some of the details. See the full version for an explanation of the differences.

The main idea is to store just a subset of the pairwise distances in the graph, from which all distances can be approximately computed efficiently. Given an undirected graph H, and a shortest path $Q \in H$, Thorup shows that for every vertex $v \in H$, there exists a set of $O(\epsilon^{-1})$ vertices on Q, called *connections*, such that the distances (called *connection lengths*) between every vertex of H and its connections on Q can be used to approximate, in $O(\epsilon^{-1})$ time, the length of any shortest path in H that intersects Q. Thorup essentially proves the following:[1]

Lemma 1. *Let Q be any shortest path in an undirected graph H. There exist sets $C(u, Q)$ of $O(\epsilon^{-1})$ vertices of Q for all $u \in H$, where:*

1. $C(u, Q)$ *are called the* connections *of u on Q.*
2. *The distance between u and a connection $q \in C(u, Q)$ is called the* connection length *of u and q.*
3. *For every $u, w \in H$, if a shortest u-to-w path in H intersects Q, then*
   $\delta_{H_Q^{uw}}(u, w) \leq (1 + \epsilon)\delta_H(u, w)$.
   Here H_Q^{uw} is the graph with vertices u, w, and the vertices of the reduction of Q to $C(u, Q) \cup C(w, Q)$, and with u-to-Q and Q-to-w edges whose lengths are the corresponding connection lengths of $C(u, Q)$ and $C(w, Q)$.

Note that, since for every v $|C(v, Q)| = O(\epsilon^{-1})$, computing $\delta_{H_Q^{uw}}(u, w)$ can be done in time that only depends on ϵ^{-1} (in fact in $O(\epsilon^{-1})$ time). Lemmas 2, 3, and 4 establish the correctness of Lemma 1.

For efficiency reasons, instead of storing exact connection lengths $\delta(\cdot, \cdot)$, the algorithm computes approximate connection lengths, which we denote by $\ell(\cdot, \cdot)$.

This following definition captures the intuitive idea that if a v-to-q path that goes through q^* is not too much longer than the shortest v-to-q path, then it suffices to store the distance from v to q^* and the distance from q^* to q.

Definition 1. q^* ϵ-covers q w.r.t. v if $\ell(v, q^*) + \delta(q^*, q) \leq (1 + \epsilon)\delta(v, q)$.

Thorup [11] uses a different notion of covering.[2]

Definition 2. q^* quasi ϵ-covers q w.r.t. v if $\ell(v, q^*) + \delta(q^*, q) \leq \delta(v, q) + \epsilon\ell(v, q^*)$.

Let Q be a path. A set C of vertices of Q is a *(quasi)-ϵ-covering* of Q w.r.t. v if for every $q \in Q$ there is a connection $q^* \in C$ that (quasi)-ϵ-covers q w.r.t. v.

A covering set is called *clean* if it is inclusion-wise minimal and *ordered* if it is sorted by the order of connections along the path Q (The endpoint of

---

[1] Thorup's treatment [11] of the undirected step does not contain the full details. See the full version of this paper for ellaboration.

[2] The term quasi-ϵ-cover is not used by Thorup. He uses ϵ-covers for this notion.

Q considered as the first vertex of Q can be arbitrarily chosen). Observe that keeping the distance of every $q \in Q$ from the first vertex of Q, allows computing $\delta_Q(q, q')$ for any $q, q' \in Q$ in constant time.

The notions of ϵ-covering sets and quasi-ϵ-covering sets are related by the following proposition:

Proposition 1. *Let $C(v, Q)$ be a quasi ϵ-covering set for some $\epsilon \in (0, \frac{1}{2})$. Then $C(v, Q)$ is a 2ϵ-covering set.*

Proof. If q^* quasi ϵ-covers q then $\ell(q^*, v) \leq \frac{1}{1-\epsilon} \delta(q, v) \leq 2\delta(q, v)$. Hence $\delta(q, q^*) + \ell(q^*, v) \leq \delta(q, v) + \epsilon \ell(q^*, v) \leq (1 + 2\epsilon)\delta(q, v)$. Therefore, if $C(v, Q)$ is a quasi ϵ-covering set, it is a 2ϵ-covering set. □

The following lemma shows that, in order to prove Lemma 1, it is suffices that the sets $C(v, Q)$ be ϵ-covering sets of size $O(\epsilon^{-1})$.

Lemma 2 (*[7, Lemma 4.1$^\beta$]*). *Let u, w be vertices in an undirected graph H. Let Q be a shortest path in H such that a u-to-w shortest path intersects Q. Let $C(u, Q), C(w, Q)$ be ϵ-covering sets of Q w.r.t. u, w, respectively. Let H_Q^{uw} be as in the statement of Lemma 1. Then,*

$$\delta_{H_Q^{uw}}(u, w) \leq (1 + \epsilon)\delta_H(u, w) \tag{1}$$

Thorup shows how to efficiently construct quasi-ϵ-covering sets. Let Q be a shortest path in an undirected graph H. Let $sssp(Q, H)$ be the smallest number s.t. for any subgraph H_0 of H, and any vertex $q \in Q_0$, where Q_0 is the reduction of Q to H_0, we can compute single source shortest paths from q in the graph $Q_0 \cup H_0$ in $O(sssp(Q, H)|E(H_0)|)$ time. It is easy to see that a standard implementation of Dijkstra's algorithm with priority queues implies $sssp(Q, H) = O(\lg |E(H)|)$. If H is planar, then $sssp(Q, H) = O(1)$ by [3].

Lemma 3 (*[11, Lemma 3.18]*). *Given an undirected graph H and shortest path Q, quasi ϵ-covering sets of Q with respect to all vertices of H, each of size $O(\epsilon^{-1} \lg n)$, can be constructed in $O(\epsilon^{-1} sssp(Q, H)|E(H)| \lg(|V(Q)|))$ time.*

By Proposition 1 the quasi ϵ-covers produced by Lemma 3 are 2ϵ-covering sets. However, their sizes are too large. The sizes can be decreased using the following thinning procedure. The proof appears in the full version.[4]

Lemma 4. *Let Q be a path in an undirected graph, and let v be a vertex. Let $D(v, Q)$ be an ordered ϵ_0-cover of Q w.r.t. v. For any $\epsilon_1 \leq 1$, a clean and ordered $(2\epsilon_0 + \epsilon_1)$-cover $C(v, Q) \subseteq D(v, Q)$ of size $O(\epsilon_1^{-1})$ can be constructed in $O(|D(v, Q)|)$ time.*

---

[3] Klein showed this lemma for ϵ-covering sets, while Thorup showed a similar lemma using a different notion of ϵ-covering sets.

[4] In [11] a thinning procedure is given only for the directed case, and it is claimed that quasi-ϵ-covering sets can be thinned. We believe this is not correct. See the full version. Instead, we give here a thinning procedure for ϵ-covering sets.

Thus, by combining Lemma 3, Proposition 1, and Lemma 4, we get the following corollary, which, along with Lemma 2, establishes Lemma 1.

Corollary 1. *Given an undirected graph H and a shortest path Q, ϵ-covering sets of Q with respect to all vertices of H, each of size $O(\epsilon^{-1})$, can be constructed in $O(\epsilon^{-1}sssp(Q,H)|E(H)|\lg(|V(Q)|))$ time.*

We now describe Thorup's distance oracle. The construction is recursive, using shortest path separators.

Lemma 5 *(Fundamental Cycle Separator [9]).* *Let H be a triangulated undirected planar graph with a rooted spanning tree T and function w assigning nonnegative weights to edges. One can find an edge $e \notin T$ such that neither the weight strictly enclosed by the fundamental cycle of e nor the weight not enclosed by the fundamental cycle of e exceeds $\frac{2}{3}$ the weight of H.*

A planar graph G can be decomposed by computing a shortest path tree for an arbitrary vertex, and applying Lemma 5 recursively. Choosing the spanning tree in Lemma 5 to be a shortest path tree guarantees that each fundamental cycle separator found consists of two shortest paths. The decomposition can be represented by a binary tree $\mathcal{T}$ in the following manner.[5]

- Each node r of $\mathcal{T}$ is associated with a subgraph G_r of G. The subgraph associated with the root of $\mathcal{T}$ is all of G.
- Each non-leaf node r of $\mathcal{T}$ is associated with S_r, the set of two shortest paths in the fundamental cycle separator found by invoking Lemma 5 on G_r.
- Each non-leaf node r has two children, whose associated subgraphs are the interior and exterior of S_r. The vertices and edges of the separator belong to both subgraphs.

Let r be a node of $\mathcal{T}$. The *frame* F_r of G_r is the set of (shortest) paths in $\bigcup_{r'}(E(S_{r'}) \cap G_r)$, where the union is over strict ancestors r' of r in $\mathcal{T}$. Each non-leaf node r stores its frame F_r. A standard argument shows that, by alternating the separation criteria between number of edges in the graph and number of paths in the frame, one can get frames consisting of a constant number of paths.

For $r \in \mathcal{T}$, let G_r° denote the subgraph of $G_r \setminus F_r$. That is, G_r° is the graph obtained from G_r by removing the edges of the frame F_r as well as any vertices of F_r that become isolated as a result of the removal. The sizes of the G_r°'s decrease by a constant factor along $\mathcal{T}$, while the sizes of the G_r's need not because there is no bound on the size of the fundamental cycle in Lemma 5. This may pose a problem, since the frame F_r is stored by every node r. To overcome this, the algorithm stores the reduction of F_r to G_r° instead of F_r itself.

Let u, w be vertices of G. Let r_u, r_w be leaves of $\mathcal{T}$ such that $u \in G_{r_u}$ and $w \in G_{r_w}$. Let r be the LCA of r_u and r_w in $\mathcal{T}$. Observe that u and w are separated by S_r. Hence, every u-to-w path in G must intersect S_r. However, a u-to-w path may or may not intersect F_r.

---

[5] We refer to the vertices of $\mathcal{T}$ as *nodes* to distinguish them from the vertices of G.

Suppose first that a shortest u-to-w path P (in G) does intersect F_r. We write $P = P_0 \circ P_1$. Path P_0 is a maximal prefix of P whose vertices belong to G_r°. We call this kind of paths *local* paths. Note that local paths start at a vertex of G_r°, end at a vertex of F_r and are confined to G_r°. Path P_1 consists of the remainder of P, and is referred to as a *global* path. Note that global paths start at a vertex of $F_r \cap G_r^\circ$, end at a vertex of G_r°, but are *not* confined to G_r°. It is not difficult to convince oneself that, to be able to approximate $\delta_G(u, w)$, it suffices to keep, for every $Q \in F_r$, *local* connections $C(u, Q)$ (i.e. the connection lengths are relative to G_r°, not the entire G) and global connections $C(w, Q)$ of (i.e. the connection lengths are relative to the entire graph G).

Now suppose that no shortest u-to-w path P (in G) intersects F_r. Then every u-to-w path P (in G) intersects S_r and is confined to G_r°. Then, to approximate P it suffices to keep, for every $Q \in S_r$, local connections $C(u, Q)$ and $C(w, Q)$.

The distance oracle therefore keeps, for each $r \in \mathcal{T}$, for each vertex $u \in G_r^\circ$:

1. local connections $C(u, Q)$ for all $Q \in F_r$.[6]
2. global connections $C(u, Q)$ for all $Q \in F_r$.
3. local connections $C(u, Q)$ for all $Q \in S_r$.

These connections, over all $u \in G_r$ and all paths in $F_r \cup S_r$ are called the (local or global) connections of r. In addition, the data structure stores:

- A mapping of each vertex $v \in G$ to a leaf node $r_v \in \mathcal{T}$ s.t. $v \in G_{r_v}$.
- A least common ancestor data structure over $\mathcal{T}$.

The space bottleneck is the size of the sets maintained. Each vertex v belongs to G_r° for $O(\lg n)$ nodes r of $\mathcal{T}$. For each of the $O(1)$ paths in the frame and separator of each such node r, v has a set of $O(\epsilon^{-1})$ connections. Hence the total space required by Thorup's oracle is $O(\epsilon^{-1} n \lg n)$.

We next describe how a query is performed. Given a u-to-w distance query, let r be the least common ancestor of r_u and r_w in $\mathcal{T}$. The algorithm computes, for each path Q of $S_r \cup F_r$ the length of a shortest u-to-w path that intersects Q using the connections $C(u, Q)$ and $C(w, Q)$ (both local and global). By construction of $\mathcal{T}$, the number of such paths Q is constant. It is easy to see that computing the distance estimate for each Q can be done in $O(\epsilon^{-1})$ time. Thus, an $(1 + \epsilon)$-approximate distance is produced in $O(\epsilon^{-1})$ time.

Efficient Construction. We now mention some, but not all the details of Thorup's $O(\epsilon^{-2} n \lg^3 n)$-time construction algorithm. Refer to [11, Subsect. 3.6] for the full details. The computation of the connections and connection lengths is done top-down the decomposition tree $\mathcal{T}$. Naively using Corollary 1 on G_r° for all $r \in \mathcal{T}$ is efficient, but only generates local connections on S_r. Using Corollary 1 on G_r would produce local connections on F_r, but is not efficient since $|F_r|$ can be much larger than $|G_r^\circ|$. Instead, For each path Q in F_r, the algorithm uses the reduction $\bar{Q}$ of Q to the vertices of Q that belong to G_r°. Let G_r^Q be the graph composed of G_r° and $\bar{Q}$. Note that $|G_r^Q| = O(|G_r^\circ|)$. The local connections on F_r can now be computed by applying Corollary 1 to G_r^Q.

---

[6] These connections are only required for the efficient construction.

It remains to compute global connections. Recall that these connection lengths reflect distances in the entire graph, not just in G_r. Clearly, applying Corollary 1 on G for every r is inefficient. Instead, the computation is done top-down $\mathcal{T}$ using an auxiliary construction. This construction augments G_r° with ϵ-covers of the separators of all ancestors of r in $\mathcal{T}$ w.r.t the vertices of G_r°. These ϵ-covers have already been computed (local connections at the ancestor), and represent distances outside G_r. Due to space constraints we defer the details to the next section, where we handle the more general case of vertex labels.

4   Undirected Approximate Vertex-Label Distance Oracle

The idea is to adapt Thorup's oracle (Sect. 3) to the vertex-label case. Thorup's oracle supports one-to-one (vertex-to-vertex) distance queries, whereas here we need one-to-many distance queries. Given two vertices u, v, Thorup's oracle finds the LCA of r_u and r_v in $\mathcal{T}$, and uses its connections to produce the answer. In a one-to-many query, there is no analogue for v. We do know, however, that a shortest u-to-λ path must intersect the separator of the leafmost node r in $\mathcal{T}$ that contains u and some λ-labeled vertex. The node r takes the role of the LCA of r_u and r_v. In order to be able to use r's connections in a distance query one must make sure that r's connections represent approximate distances to λ-labeled vertices in the entire graph, not just in G_r°.

We define a set $\mathcal{L}$ of new (artificial) vertices, one per label. For every $r \in \mathcal{T}$, let $\mathcal{L}_r = \{\lambda \in \mathcal{L} | G_r^\circ \cap V_\lambda \neq \emptyset\}$ be the restriction of $\mathcal{L}$ to labels in G_r°.

Simply connecting each vertex of V_λ to an artificial vertex representing the label λ is bound to fail. To see why, suppose vertices u and v both have label λ. Adding an artificial vertex λ and zero-length undirected edges $v\lambda$ and $u\lambda$ creates a zero-length path between u and v that does not exist in the original graph. While this does not change the distance between any vertex and its closest λ-labeled vertex, it may change distances between a vertex and its closest λ'-labeled vertex ($\lambda' \neq \lambda$). Therefore, we would have liked to add, for each label λ *separately*, a single artificial vertex λ, and compute the connection sets $C(\lambda, Q)$. Doing so would result in correct distance estimates, but is not efficient. We show how to compute the connections $C(\lambda, Q)$ without actually performing this inefficient procedure. Instead of having a single artificial vertex per label, it is split into many artificial vertices (one for each incident edge). The problem with this approach is that the number of connections becomes too large (each split vertex has its own set of $O(\epsilon^{-1})$ connections). We use an extension of the thinning procedure (Lemma 4) to select a small subset of these connections and still get the desired approximation.

Another point that we must address is that, for $\lambda \in \mathcal{L}_r$, the global connections $C(\lambda, Q)$ should reflect the minimum distances between the connections of λ on Q to the closest vertex with label λ in G, not just to vertices with label λ in G_r°. We show how to achieve this by an extension of the auxiliary construction used to compute the global connections in Thorup's unlabeled oracle. We start with the extended thinning lemma.

Lemma 6 *Let $\{u_i\}$ be vertices and Q be a shortest path. Given ordered ϵ-covering sets $\{D(u_i, Q)\}$ it is possible to compute in linear time a clean and ordered 3ϵ-covering connections set C of size $O(\epsilon^{-1})$ which represent approximated distances from any $q \in Q$ to its closest vertex among $\{u_i\}$.*

Proof We first convert every connection length $\ell(u_i, q)$ for every q in $D(u_i, Q)$ to reflect an approximated length from q to its closest vertex $u^* \in \{u_j\}$, rather than to u_i. We obtain these lengths using the fact that q is ϵ-covered with respect to u^* by some connection in $D(u^*, Q)$. Let Z_u be the graph composed of:

1. $\bar{Q}$, the reduced form of Q to connections of all $\{D(u_i, Q)\}$.
2. vertices $\{u_i\}$, along with edges between each u_i to its connections, with lengths equal to the corresponding connection lengths.
3. vertex u, connected with zero-length edges to all $\{u_i\}$.

By the ϵ-covering property, the distances between every $q \in \bar{Q}$ and u in Z_u represent approximate distances between q and its closest vertex $u^* \in \{u_j\}$ in G. To see this, assume $q \in \bar{Q}$ is a connection of u_1, and is closest to u^*. Let q^* be a connection of u^* which ϵ-covers q w.r.t. u^*. Then $\delta_{Z_u}(q, u^*) \leq \delta_Q(q, q^*) + \ell(q^*, u^*) = \delta_G(q, q^*) + \ell(q^*, u^*) \leq (1 + \epsilon)\delta_G(q, u^*)$.

All shortest paths from u in Z_u can be computed in linear time; first, relax all edges incident to u and $\{u_i\}$. Then, relax the edges of $\bar{Q}$ by going first in one direction along Q and then relaxing the same edges again in the other direction. For connection p on $\bar{Q}$, a u-to-p shortest path first reaches Q along one of $\{u_i\}$ edges and then walks along Q toward p. Hence the relaxation was done in the correct order. We update the connection lengths to the distances thus computed.

Let $\tilde{D}(u, Q)$ denote the ordered union of all connections, along with the updated connection lengths. Since all $\{D(u_i, Q)\}$ were ordered, it is possible to order their union in linear time. Let G_u be the graph obtained from G by adding an apex u connected with zero length edges to all $\{u_i\}$. We stress that G_u is not constructed by the algorithm, but only used in the proof. $\tilde{D}(u, Q)$ is an ϵ-cover of Q with respect to u in G_u. Now apply Lemma 4 to $\tilde{D}(v, Q)$ with $\epsilon_0 = \epsilon_1 = \epsilon$ to obtain a 3ϵ-cover of Q with respect to u in G_u of size $O(\epsilon^{-1})$. $\qquad\square$

Vertex-Label Distance Oracle for Undirected Graphs. The vertex labeled distance oracle is very similar to the unlabeled one (Sect. 3). It uses the same decomposition tree $\mathcal{T}$, and stores, for each $r \in \mathcal{T}$, the same covering sets. The only difference is that, in addition to the covering sets $C(u, Q)$ for each vertex $u \in G_r^\circ$, the oracle also stores connection information for labels as we now explain.

For every $r \in \mathcal{T}$ and $\lambda \in \mathcal{L}_r$, the oracle stores local and global connections $C(\lambda, Q)$. The local connections $C(\lambda, Q)$ are connections in the graph obtained from G_r° by adding an artificial vertex λ, along with zero length edges from all λ-labeled vertices in G_r° to λ. The global connections $C(\lambda, Q)$ are connections in the graph obtained from G by adding an artificial vertex λ, along with zero length edges from all λ-labeled vertices in G to λ. Before explaining how to compute these connections we discuss how a distance query is performed.

Obtaining the distance from u to λ is done by finding the lowest ancestor r of r_u with $\lambda \in \mathcal{L}_r$. A shortest u-to-λ path must cross S_r, and perhaps also F_r. The algorithm estimates, for each path Q of $S_r \cup F_r$, the length of a shortest u-to-λ path that intersects Q, using the connections $C(u, Q)$ and $C(\lambda, Q)$ stored for r (Since $\lambda \in \mathcal{L}_r$, r does store Q-to-λ connections).

Finding r can be done by binary search on the path from r_u to the root of $\mathcal{T}$. The number of steps of the binay search is $O(\lg \lg n)$. Finding whether a node r' has a vertex with label λ can be done, e.g., by storing all unique labels in $G_{r'}^\circ$ in a binary search tree, or by hashing. In the former case finding r takes $O(\lg \lg n \lg |L|)$ time, and in the latter $O(\lg \lg n)$, (in the word-RAM model).

It remains to show how the connections are computed. We begin with the local connections. For every $r \in \mathcal{T}$, for every $Q \in F_r \cup S_r$, the algorithm computes ordered ϵ-covering sets of connections on Q w.r.t. each vertex of G_r° to Q by invoking Corollary 1 to G_r°. This takes $O(\epsilon^{-1}|V(G_r^\circ)| \lg n)$ time (using [3] for shortest path computation). For each $\lambda \in \mathcal{L}_r$, let n_λ denote the number of λ-labeled vertices in G_r°. The total number of connections to λ-labeled vertices in G_r° is $O(\epsilon^{-1}n_\lambda)$. The algorithm next applies the extended thinning lemma (Lemma 6) to get a connections set $C(\lambda, Q)$ of size $O(\epsilon^{-1})$ in $O(\epsilon^{-1}n_\lambda)$ time. Since $\sum_\lambda n_\lambda = O(|V(G_r^\circ)|)$, the runtime for a single r and Q is $O(\epsilon^{-1}|V(G_r^\circ)|)$.

We now show how to compute the global connections without invoking Corollary 1 to the entire input graph G at every call.

Lemma 7. *Let $r \in \mathcal{T}$. Global connections for r can be computed using just the (local) connections of strict ancestors of r. Computing all global connections for all $r \in \mathcal{T}$ can be done in $O(\epsilon^{-2}n \lg^3 n)$ time.*

Proof. Let Q be a path in F_r. Let X_r^Q be the graph composed of the following:

- The vertices $\mathcal{L}_r$
- The vertices and edges of $\bar{Q}$, the reduction of Q to $V(Q) \cap V(G_r^\circ)$.
- For each strict ancestor r' of r, for each path $Q' \in S_{r'}$, the vertices and edges of $\bar{Q}'$, the reduction of Q' to vertices that are (local) connections (in $G_{r'}^\circ$) of Q' w.r.t. vertices in $Q \cup \mathcal{L}_r$, along with edges representing the corresponding connection lengths.

The algorithm creates a graph $\hat{X}_r^Q$ from X_r^Q by breaking every artificial vertex λ in X_r^Q into many copies $\{\lambda_e\}$, one per incident edge of λ. We stress that the artificial vertices λ_e are not directly connected to each other in $\hat{X}_r^Q$. Hence, the problem of shortcuts mentioned earlier is avoided.

Note that splitting vertices in this way does not increase the number of edges in the $\hat{X}_r^Q$. The algorithm applies Corollary 1 to $\hat{X}_r^Q$ and Q, obtaining a small sized ϵ-cover $C(\lambda_e, Q)$ for every λ_e.

Let q be any vertex of $\bar{Q}$, and let λ be a label in G_r°. Let P be a shortest q-to-λ path in G. Let r' be the rootmost strict ancestor of r such that $S_{r'}$ is intersected by P. Note that r' must exist since $q \in F_r$, so q belongs to the separator of some strict ancestor of r. Thus P is entirely contained in $G_{r'}^\circ$. Let Q' be a path in $S_{r'}$ intersected by P. By construction of $\hat{X}_r^Q$, it contains an ϵ-covering set of connections of Q' with respect to q in $G^{\circ}_{r'}$, as well as the edges of $\bar{Q}'$ and an ϵ-covering set of connections of Q' with respect to λ in $G^{\circ}_{r'}$. Hence, by Lemma 2, there exists a shortest q-to-λ_e path (for some artificial vertex λ_e) in $\hat{X}^Q_r$ whose length is at most $(1 + \epsilon)$ times the length of P. On the other hand, because the vertices λ_e (for any $\lambda \in \mathcal{L}_r$) are not directly connected to each other in $\hat{X}^Q_r$, every path in $\hat{X}^Q_r$ corresponds to some path in G, so shortest paths in $\hat{X}^Q_r$ are at least as long as those in G. This proves that $\hat{X}^Q_r$ correctly represents all desired global connection lengths.

We proceed with describing the construction of the connection sets of the appropriate sizes. To bound the size of the connections $\{C(\lambda_e, Q)\}$, we count the number of edges incident to λ in X^Q_r (i.e., before it is split). There is an edge for each of the $O(\epsilon^{-1})$ connections of λ on each of the $O(\lg n)$ paths of separators of ancestors of r. For each such edge there is a vertex λ_e with an ϵ-covering set of $\bar{Q}$ of size $O(\epsilon^{-1})$. Thus, the total number of connections of $\bar{Q}$ for all λ_e vertices is $O(\epsilon^{-2} \lg n)$. The algorithm applies Lemma 6, the extended thinning procedure, to $\{C(\lambda_e, Q)\}_e$ to get $C(\lambda, Q)$ of size $O(\epsilon^{-1})$. Doing so for all labels in G°_r requires $O(\epsilon^{-2} \lg n + \epsilon^{-1}|\mathcal{L}_r|)$ space.

We now bound the running time. Since splitting vertices does not increase the number of edges, applying Corollary 1 to $\hat{X}^Q_r$ takes $O(\epsilon^{-2}|V(G^{\circ}_r)| \lg^2 n)$ time. Applying Lemma 6 is done within the same time bound. To conclude, the total runtime over all nodes of $\mathcal{T}$ is $O(\epsilon^{-2} n \lg^3 n)$. $\square$

References

1. Abraham, I., Chechik, S., Krauthgamer, R., Wieder, U.: Approximate nearest neighbor search in metrics of planar graphs. APPROX/RANDOM **2015**, 20–42 (2015)
2. Chechik, S.: Improved distance oracles and spanners for vertex-labeled graphs. In: Epstein, L., Ferragina, P. (eds.) ESA 2012. LNCS, vol. 7501, pp. 325–336. Springer, Heidelberg (2012)
3. Henzinger, M.R., Klein, P.N., Rao, S., Subramanian, S.: Faster shortest-path algorithms for planar graphs. J. Comput. Syst. Sci. **55**(1), 3–23 (1997)
4. Hermelin, D., Levy, A., Weimann, O., Yuster, R.: Distance oracles for vertex-labeled graphs. In: Aceto, L., Henzinger, M., Sgall, J. (eds.) ICALP 2011, Part II. LNCS, vol. 6756, pp. 490–501. Springer, Heidelberg (2011)
5. Kawarabayashi, K., Klein, P.N., Sommer, C.: Linear-space approximate distance oracles for planar, bounded-genus and minor-free graphs. In: Aceto, L., Henzinger, M., Sgall, J. (eds.) ICALP 2011, Part I. LNCS, vol. 6755, pp. 135–146. Springer, Heidelberg (2011)
6. Kawarabayashi, K., Sommer, C., Thorup, M.: More compact oracles for approximate distances in undirected planar graphs. In: SODA 2013, pp. 550–563 (2013)
7. Klein, P.N.: Preprocessing an undirected planar network to enable fast approximate distance queries. In: SODA 2002, pp. 820–827 (2002)
8. Li, M., Ma, C.C.C., Ning, L.: $(1 + \epsilon)$-distance oracles for vertex-labeled planar graphs. In: Chan, T.-H.H., Lau, L.C., Trevisan, L. (eds.) TAMC 2013. LNCS, vol. 7876, pp. 42–51. Springer, Heidelberg (2013)
9. Lipton, R., Tarjan, R.: A separator theorem for planar graphs. SIAM J. Appl. Math. **36**, 177–189 (1979)

10. Łącki, J., Ocwieja, J., Pilipczuk, M., Sankowski, P., Zych, A.: The power of dynamic distance oracles: efficient dynamic algorithms for the steiner tree. In: STOC 2015, pp. 11–20 (2015)
11. Thorup, M.: Compact oracles for reachability and approximate distances in planar digraphs. J. ACM **51**(6), 993–1024 (2004)

Constant-Time Local Computation Algorithms

Yishay Mansour[1,2], Boaz Patt-Shamir[1], and Shai Vardi[1(✉)]

[1] Tel Aviv University, Tel Aviv, Israel
{mansour,shaivar1}@post.tau.ac.il, boaz@tau.ac.il
[2] Microsoft Research, Herzliya, Israel

Abstract. Local computation algorithms (LCAs) produce small parts of a single solution to a given search problem using time and space sublinear in the size of the input. In this work we present LCAs whose time complexity (and usually also space complexity) is *independent of the input size*. Specifically, we give (1) a $(1 - \epsilon)$-approximation LCA to the maximal weighted base of a graphic matroid (i.e., maximal acyclic edge set), (2) LCAs for approximating multicut and integer multicommodity flow on trees, and (3) a local reduction of weighted matching to any unweighted matching LCA, such that the running time of the weighted matching LCA is also independent of the edge weight function.

1 Introduction

Local computation algorithms (LCAs) provide a solution to situations in which we require fast and space-efficient access to part of a solution to a large computational problem, but we never need the entire solution at once. Consider, for instance, a database describing a network with millions of nodes and edges, on which we would like to compute a maximal matching. At any point in time, the algorithm may be queried about an edge, and is expected to reply "yes" or "no", depending on whether the edge part of a maximal matching. The algorithm may never be required to compute the entire solution. However, replies to queries are expected to be consistent with a single matching.

LCA Measures. Typically, LCAs use polylog(n) space, and are required to reply to each query in polylog(n) time. [1] Some papers on LCAs (e.g., [12]) give three criteria for measuring LCAs: running time per query, the total space required, and failure probability. [2] Others (e.g., [2]), consider only the number of times

Y. Mansour—Supported in part by a grant from the Israel Science Foundation, by a grant from United States-Israel Binational Science Foundation (BSF), by a grant from the Israeli Ministry of Science (MoS) and the Israeli Centers of Research Excellence (I-CORE) program (Center No. 4/11).

B. Patt-Shamir—Supported in part by the Israel Science Foundation (grant No. 1444/14) and by the Israel Ministry of Science and Technology.

S. Vardi—Supported in part by the Google Europe Fellowship in Game Theory.

[1] We assume the standard uniform-cost RAM model, in which the word size is $O(\log n)$ bits, where n is the input size.

[2] The failure probability of an LCA A (cf. [7]) is the probability, taken over coin flips of A, that the running time of A for any query exceeds its stated running time (and not the probability that A errs).

© Springer International Publishing Switzerland 2015
L. Sanità and M. Skutella (Eds.): WAOA 2015, LNCS 9499, pp. 110–121, 2015.
DOI: 10.1007/978-3-319-28684-6_10

the input is probed, and the amount of information the LCA needs to store between queries. In this paper we propose a more comprehensive model that unifies the two approaches. The idea is to distinguish between computational time and probe complexities, and between enduring and transient memory. More specifically, a *probe* to the input graph consists of asking a vertex for a list of its neighbors.[3] The point is that the cost of a probe may sometimes be much higher than the cost of a computational step. Regarding space, we assume that there is an *enduring memory*, which is written only once by the algorithm, before the first query is presented. We think about it as an augmentation of the input. Enduring memory is useful in randomized LCAs (e.g., [1,7,12]), where the LCA must use the same randomness each time it is invoked to ensure consistency. This can be done by storing a random seed in the enduring memory. *Transient memory* is simply the memory required to compute a reply to each query. Note that in our formulation, the algorithm's reply to a query depends only on the input and the enduring memory, and not on the history. Formal definitions are provided in Sect. 2.

New LCAs. In this paper we give constant-time, constant-probe LCAs to the following graph problems, assuming graphs with constant maximal degree.

- *Graphic Matroids.* Given a weighted graph, the task is to find an acyclic edge set (forest) of approximately the maximum possible weight. In the corresponding LCA, a query specifies an edge, and the algorithm says whether the given edge is in the solution forest. We present a deterministic $(1 - \epsilon)$-approximate LCA for graphic matroids, whose running time and space are independent of the size of the matroid (see Sect. 2 for formal definitions).
- *Integer Multi-commodity Flow (IMCF) and Multicut on Trees.* Given a tree with capacitated edges and source-destination pairs, the goal of IMCF is to route the greatest possible total flow where each pair represents a different commodity, subject to edge capacity constraints. Multicut is the dual problem where the goal is to pick an edge set of minimal total capacity so that no source can be connected to its destination without using a selected edge. We give a deterministic LCA for IMCF and multicut on trees that runs in constant time and gives a $(1/4)$-approximation to the optimal IMCF and a 4-approximation to the minimum multicut. We also give a randomized LCA to IMCF, with constant running time, very little enduring memory (less than a word), and expected approximation ratio $\frac{1}{2} - \epsilon$ for any constant $\epsilon > 0$.
- *Weighted Matching.* Given a weighted graph, we would like to approximate the maximum weight matching. We design a deterministic reduction from any (possibly randomized) LCA A for unweighted matching with approximation ratio α to weighted matching with approximation ratio $\alpha/8$. Our reduction invokes A a constant number of times. Both the running time and approximation ratio are independent of the magnitude of the edge weights.

Related Work. LCAs were introduced by Rubinfeld et al. [12]. Alon et al. [1] described LCAs for hypergraph 2-coloring and maximal independent set (MIS)

---

[3] We typically assume that vertex degrees are bounded by a constant.

on graphs of bounded degree, using a reduction from parallel and distributed algorithms. Mansour et al. [6], extending results of [1], showed how to convert a large class of online algorithms to LCAs using the technique of Nguyen and Onak [9], in which a random ordering is generated over the vertices, and this ordering is used to simulate the online algorithm: in order to reply to a query about a vertex, we need to simulate the online algorithm on all vertices that come before it in the ordering. The main challenge is to bound the number of vertices that one needs to probe per query. Recently, Reingold and Vardi [11] extended these results to a much wider class of graphs, and obtained stronger bounds.

For graphs of bounded degree, Even et al. [2] showed how to obtain acyclic orientations of the edges using the distributed coloring algorithms of Linial [4] and Panconesi and Rizzi [10]. They were thus able to obtain deterministic LCAs with running time dependent on $\log^* n$. They also give algorithms with similar running times for approximate maximum cardinality (MCM) and maximum weight matchings (MWM). Their results for MWM depend logarithmically on the ratio of the maximum to the minimum weight.[4]

The reader may also find it interesting to compare the results of this paper with other algorithms that run in constant time, such as distributed constant-time algorithms (see [13] for a recent survey), or constant-time approximation algorithms (e.g., [9,18]).

We give more problem-specific related work in the relevant sections.

2   Preliminaries

General Concepts. We denote the set $\{1, 2, \ldots n\}$ by $[n]$.

Graph Concepts. Let $G = (V, E)$ be a simple undirected graph. The neighborhood of a vertex v, denoted $N(v)$, is the set of vertices that share an edge with v: $N(v) = \{u : (u, v) \in E\}$. The *degree* of a vertex v, is $|N(v)|$. The *distance* between two vertices u and v, denoted $\text{dist}(u, v)$, is the minimal number of edges required to reach one from the other. For any vertex v, its k-neighborhood, denoted $N^k(v)$ is the set of all vertices at distance at most k from v. (Note that $N(v) = N^1(v) \setminus N^0(v)$.) For any edge $e = (u, v)$, its k-neighborhood is defined as $N^k(e) = N^k(u) \cup N^k(v)$. Given a non-empty vertex set $S \neq V$, a *cut* $(S, V \setminus S)$ is the set of edges with exactly one endpoint in S.

Throughout this paper, we assume that the maximal degree in G is upper bounded by some constant parameter d. For simplicity of presentation, we assume that G is d-regular and that $|V|$ and d are powers of 2. All our results hold without these assumptions.

---

[4] We note that while our algorithm for MWM runs in constant time, independently of the size of the graph and of the edge weights, its approximation guarantee is much worse than that of [2], whose approximation factor is $(1 - \epsilon)$.

Approximation Algorithms. We define approximation algorithms as follows.

Definition 1. *Given a maximization problem over graphs and a real number $0 \leq \alpha \leq 1$, a (possibly randomized) α-approximation algorithm $\mathcal{A}$ is guaranteed, for any input graph G, to output a feasible solution whose (possibly expected) value is at least an α fraction of the value of an optimal solution.[5] The definition of approximation algorithms to minimization problems is analogous, with $\alpha \geq 1$.*

LCAs. We extend the model of [12] for local computation algorithms (LCAs) to distinguish between time and probe complexity, and between enduring and transient memory.

Definition 2 (LCA). *A $(t(n), p(n), em(n), tm(n), \delta(n))$-local computation algorithm $\mathcal{A}$ for a computational problem is a (possibly randomized) algorithm that receives an input of size n and a query x. Before the first query, $\mathcal{A}$ is allowed to write $em(n)$ bits to the enduring memory, and may only read from it thereafter. Algorithm $\mathcal{A}$ makes at most $p(n)$ probes to the input in order to reply to any query x, and does so in time $t(n)$ using $tm(n)$ bits of transient memory (in addition to the enduring memory). The probability that $\mathcal{A}$ deviates from the probe, time or transient memory bounds[6] (i.e., uses more than the prescribed amounts) is at most $\delta(n)$, which is called $\mathcal{A}$'s failure probability. Algorithm $\mathcal{A}$ must be consistent, that is, the algorithm's replies to all possible queries conform to a single feasible solution to the problem.*

Remark 3. All LCAs of this paper have failure probability 0; we include it in the model for compatibility with previous models, but omit it from the statements for brevity. Furthermore, all the algorithms presented in this paper have a constant running time, and hence the transient memory is guaranteed to be of constant size; we omit this measure from the statements as well.

3   Graphic Matroids

In this section we consider the problem of finding the maximum weight basis of a *graphic matroid*, defined as follows.

Definition 4 (Graphic Matroid). *A graphic matroid is a matroid whose independent sets are forests in an undirected graph.*

We are given a graphic matroid $\mathcal{M} = (\mathcal{E}, \mathcal{I})$, and would like to find an independent set of (approximately) maximal weight. In graph terminology, we are given a graph $G = (V, E)$, with non-negative edge weights, $w : E \to \mathbb{R}^+$; we would like to find an acyclic set of edges of (approximately) maximal weight. That is, we seek a forest whose weight is close to the weight of a maximal spanning tree (MaxST). Without loss of generality, we assume edge weights are distinct, as

---

[5] In case of a randomized algorithm, expectation is over its random choices.
[6] Note that the LCA is *not* allowed to deviate from the enduring memory bound.

it is always possible to break ties by ID Recall that when the edge weights are distinct, there is a unique MaxST.

We first describe a parallel algorithm for finding a MaxST in a graph, and then explain how to adapt it to an LCA. We note that our parallel algorithm is less efficient than others (say, the Borůvka's algorithm [8]), if it is used as a parallel algorithm. In fact, there is no instance on which it would out-perform Borůvka's algorithm. Nevertheless, it is useful as adapting it to an LCA and analyzing its local properties are simple.

We first need a few definitions. Define the distance between a vertex v and an edge $e = (u, w)$, denoted $\text{dist}(v, e)$, to be $\min\{\text{dist}(v, u), \text{dist}(v, w)\}$.

Definition 5 (Connected Component, Truncated CC). *Let $G = (V, E)$ be a simple undirected graph. For a vertex $v \in V$, and a subset of edges $S \subseteq E$, the connected component of v with respect to S is $\text{CC}_S(v) \subseteq S$ which includes the edges $e \in S$ that have a path from e to v using only edges in S. (Note that $\text{CC}_S(\cdot)$ is a partition of S.) The k-truncated connected component of v is the set of all vertices in the connected component of v at distance at most k from v (w.r.t. G). We denote it by $k\,\text{TCC}_S(v)$.*

Algorithm 1 works as follows. We maintain a forest S, initially empty. For any vertex v, denote $\Gamma_v^k = k\,\text{TCC}_S(v)$. In round k, vertex v considers the cut $(\Gamma_v^k, V \setminus \Gamma_v^k)$ and adds the heaviest edge of the cut, say e, to S. (Note that k is both the round number and the radius parameter of the truncated connected component.) In contrast to many MaxST algorithms (such as Prim's and Kruskal's algorithms), an edge can be considered more than once, and it is possible that an edge e is considered—and even added—when it is already in S (if we add e to S when $e \in S$, S remains the same).

Algorithm 1. Parallel (CREW) MaxST Approximation Algorithm.

Input : $G = (V, E)$ with weight function $w : E \to \mathbb{R}^+, \epsilon > 0$
Output: a forest S
//assume all edge weights are distinct

For all $v \in V$, $S(v) = \emptyset$;
for *round $k = 0$ to $1/\epsilon$* **do**
⎢   For each vertex v, let $\Gamma_v^k = k\,\text{TCC}_S(v)$;
⎢   **for** *all vertices $v \in V$ in parallel* **do**
⎢ ⎢   **if** *e is the heaviest edge of the cut $(\Gamma_v^k, V \setminus \Gamma_v^k)$* **then**
⎢ ⎢ ⎣ $S(v) = S(v) \cup \{e\}$;
Return $S = \bigcup_{v \in V} S(v)$.

Correctness of Algorithm 1. The correctness of Algorithm 1 relies on the so-called "blue rule" [14].

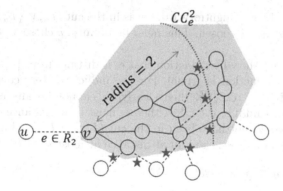

Fig. 1. The situation considered in the proof of Proposition 8, for $k = 2$. The edge $e = (u, v)$ is in $\mathcal{R}_k$. Solid edges are in $\mathcal{S}$, dashed edges are in $E \setminus \mathcal{S}$. The shaded area represents CC_e^2, and the dotted red arc represents the distance 2 range. Edges marked by $\star$ are considered by v in round 2 (Color figure online).

Lemma 6 ([14]). *Let C be any cut of the graph. Then the heaviest edge in C belongs to* MaxST.

Corollary 7. *All edges added to $\mathcal{S}$ by Algorithm 1 are in* MaxST.

Corollary 7 establishes the correctness of Algorithm 1.

Approximation Ratio. We now turn to analyze the approximation ratio of Algorithm 1. To this end, define $\mathcal{S}_k$ to be the set of edges of MaxST that were added to $\mathcal{S}$ in rounds $1, 2, \ldots, k$ (implying that $\mathcal{S}_k \subseteq \mathcal{S}_{k+1}$). Let $\mathcal{R}_k = \text{MaxST} \setminus \mathcal{S}_k$ (see Fig. 1).

Consider the *component tree* of MaxST, defined as follows: the node set is $\{CC_{\mathcal{S}_k}(v) \mid v \in V\}$, and the edge set is $\{(CC_{\mathcal{S}_k}(v), CC_{\mathcal{S}_k}(u)) \mid (v, u) \in \mathcal{R}_k\}$. In words, there is a node in the component tree for each connected component of $\mathcal{S}_k$, and there is an edge in the component tree iff there is an edge in $\mathcal{R}_k$ connecting nodes in the corresponding components. We choose an arbitrary component as the root of the component tree, and direct all the edges towards it; this way, each edge $e \in \mathcal{R}_k$ is outgoing from exactly one connected component of $\mathcal{S}_k$. We denote this component by CC_e^k. Note that $CC_e^i \subseteq CC_e^j$ for $i < j$ because components only grow. For any set of edges S, let $w(S) = \sum_{e \in S} w(e)$.

The following proposition is the key to the analysis.

Proposition 8. *For any $k \geq 1$, $\forall e \in \mathcal{R}_k$, $w(e) \leq \frac{w(CC_e^k)}{k}$.*

Proof. If $\mathcal{R}_k$ is empty, the claim holds trivially. Let $e = (v, u)$ be any edge in $\mathcal{R}_k$ (directed from v to u). Edge e was not chosen by vertex v in rounds $1, \ldots, k$. For $i \in [k]$, let e_i be the edge chosen by v in round i. It suffices to show that (1) all edges e_i are heavier than e, i.e. $\forall i \in [k], w(e_i) > w(e)$, and that (2) the edges e_i are distinct, i.e., $\forall i, j \in [k], i \neq j \Rightarrow e_i \neq e_j$.

The proof of (1) is straightforward: e was in the cut $(\Gamma_v^i, V \setminus \Gamma_v^i)$ in all rounds $i \in [k]$, but it was never chosen. This must be because v chose a heavier edge in each round.

To prove (2), we show by induction that e_i is distinct from $\{e_1, e_2, \ldots, e_{i-1}\}$. The base of the induction is trivial. For the inductive step, consider the two possible cases. If $e_i \notin CC_e^{i-1}$, then clearly, e_i cannot be any edge that was previously added. And if $e_i \in CC_e^{i-1}$, then e_i must be at distance exactly i from v, otherwise it would not have been in the cut $(\Gamma_v^i, V \setminus \Gamma_v^i)$ and could not have been added. But $e_1, \ldots, e_{i-1}$ are all at distance at most $i-1$ (w.r.t. G). □

Corollary 9. *For $k \geq 0$, $w(\mathcal{R}_k) \leq \frac{w(\mathcal{S}_k)}{k}$.*

Proof.

$$w(\mathcal{R}_k) = \sum_{e \in \mathcal{R}_k} w(e)$$

$$\leq \sum_{e \in \mathcal{R}_k} \frac{w(CC_e^k)}{k} \qquad \text{by Prop. 8}$$

$$\leq \frac{w(\mathcal{S}_k)}{k}. \qquad e \neq e' \Rightarrow CC_e^k \neq CC_{e'}^k, \text{ and } \bigcup_{e \in \mathcal{R}_k} CC_e^k \subseteq \mathcal{S}_k$$

□

This enables us to prove our approximation bound. Denote the weight of MaxST by OPT.

Lemma 10. $w(\mathcal{S}_k) \geq (1 - \frac{1}{k+1})OPT$.

Proof. As $\mathcal{R}_k = \text{MaxST} \setminus \mathcal{S}_k$,

$$\frac{w(\mathcal{S}_k)}{\text{OPT}} = \frac{w(\mathcal{S}_k)}{w(\mathcal{S}_k) + w(\mathcal{R}_k)} \geq \frac{w(\mathcal{S}_k)}{w(\mathcal{S}_k) + w(\mathcal{S}_k)/k} = \frac{k}{k+1},$$

where the inequality is due to Corollary 9. □

This concludes the analysis of Algorithm 1. We now describe the LCA we derive from it.

Adaptation to an LCA and Complexity Analysis. Given a graph $G = (V, E)$ and a query $e = (u, v) \in E$, the implementation of Algorithm 1 as an LCA is as follows. Consider iteration k of Algorithm 1. Probe G to discover $N^{2k}(u)$ and $N^{2k}(v)$. Simulate Algorithm 1 on all vertices in $N^k(u) \cup N^k(v)$ for k rounds. In each round i, for each node $s \in N^k(u) \cup N^k(v)$, the algorithm computes $\Gamma_v^i = i\,\text{TCC}_\mathcal{S}(v)$, finds the heaviest edge e in the cut $(\Gamma_v^i, V \setminus \Gamma_v^i)$, and adds it to the solution. This gives the following lemma.

Lemma 11. *The time required to simulate the execution of Algorithm 1 for k rounds as an LCA is $kd^{O(k)}$, and the probe complexity is $d^{O(k)}$.*

Proof. The time to discover $N^{2k}(v) \cup N^{2k}(u)$ by probing the graph is bounded by $d^{O(k)}$. Each vertex in $z \in N^k(v) \cup N^k(u)$ executes Algorithm 1 k rounds: in round j, it constructs S_z, by exploring $N^j(z) \subseteq N^{2k}(v) \cup N^{2k}(u)$. Overall, the time complexity is

$$d^{O(k)} + \sum_{z \in N^k(v) \cup N^k(u)} k|N^k(z)| = kd^{O(k)}.$$

$\square$

Combining Lemmas 10 and 11 gives our first main result.

Theorem 12. *There exists a deterministic LCA, that for any graph G whose degree is bounded by d and every $\epsilon > 0$, computes a forest whose weight is a $(1-\epsilon)$-approximation to the maximal spanning tree of G in time $t(n) = \frac{1}{\epsilon} d^{O(1/\epsilon)}$, probe complexity $p(n) = d^{O(1/\epsilon)}$, and enduring memory $em(n) = 0$.*

4 Multicut and Integer Multicommodity Flow in Trees

In this section we consider the integer multicommodity flow (IMCF) and multicut problems in trees. While simple, our LCAs demonstrate how, under some circumstances, one can find constant-time LCAs for apparently global problems.

The input is an undirected graph $G = (V, E)$ with a positive integer capacity $c(e)$ for each $e \in E$, and a set of pairs of vertices $\{(s_1, t_1), \ldots, (s_k, t_k)\}$. (The pairs are distinct, but the vertices are not necessarily distinct.)

In the *Integer Muticommodity Flow Problem*, the goal is to route commodity i from s_i to t_i so as to maximize the sum of the commodities routed, subject to edge capacity constraints. (There is no a priori upper or lower bound on the amount of flow for each commodity.) Note that in a tree, the only question is how much to route: the route is uniquely determined anyway. In the dual *Multicut Problem*, the goal is to find a minimum capacity *multicut*, where a multicut is an edge set that separates s_i from t_i for all $1 \leq i \leq k$.

We make the following assumptions about the input to allow for appropriately bound the time and space complexity of the algorithms. First, we assume that in the given tree each node has at most $d = O(1)$ children, and that T is rooted in the sense that the depth of each vertex is known, and is part of the properties that are found by querying the vertex. Second, we assume that the distances from s_i to t_i are bounded by some given parameter ℓ; i.e., $\forall i, \text{dist}(s_i, t_i) \leq \ell$. Our bounds will be a function of ℓ, so that if ℓ is independent of tree size, then so are the time and space complexity of our algorithms.

In the local version of IMCF (multicut), we are queried on an edge, and are required to output how much of each resource is routed on it (whether it is part of the cut). As before, we adapt a classical algorithm to an LCA. This time we use the known algorithm of Garg et al. [3] as a subroutine (denoted Algorithm 2, and presented below for completeness). Note that this is a primal-dual algorithm, that solves IMCF and multicut simultaneously.

Theorem 13 [16]. *Algorithm 2 achieves approximation factors of 2 for the multicut problem and 1/2 for the IMCF problem on trees.*

Algorithm 2. Multicut and IMCF in trees [3,16]

Input : A rooted tree T
Output: a flow f and a cut D

Initialize $f = 0, D = \emptyset$;
for *each vertex v in nonincreasing order of depth* **do**
  for *each pair (s_i, t_i) s.t. lowest common ancestor$(s_i, t_i) = v$,* **do**
    Greedily route flow from s_i to t_i if possible;
    Add all saturated edges to D in arbitrary order;

Let $e_1, \ldots, e_k$ be the ordered list of edges in D;
for $j = k$ *down to* 1 **do**
  If $D - \{e_j\}$ is a multicut, then remove e_j from D;

Our deterministic LCA finds a $(4 + \epsilon)$-approximation to the multicut problem, and in trees with minimum capacity $c_{\min} \geq 2$, the same algorithm finds an IMCF with approximation factor $\frac{\lfloor c_{\min}/2 \rfloor}{2c_{\min}} \geq \frac{1}{6}$.[7] We also present a randomized Algorithm, that gives an approximation factor of $(\frac{1}{2} - \epsilon)$ to IMCF for any desired $\epsilon > 0$ (the running time depends on $1/\epsilon$). The algorithms are similar, in that they partition the tree to subtrees and apply Algorithm 2 to each subtree. The randomized algorithm requires a very small amount of enduring memory, namely $O(\log(\ell/\epsilon))$ bits.

4.1 Deterministic LCA

We first describe the deterministic LCA. An edge is said to be at depth z if it connects vertices at depths $z - 1$ and z. The deterministic algorithm (whose pseudocode appears in the full version) for multicut is as follows. We consider two overlapping decompositions of the tree into subtrees of height 2ℓ: the first decomposition is obtained by removing all edges at depth $k\ell$ for *even* values of k, and the second is by removing all edges at depth $k\ell$ for *odd* values of k. Now, given an edge e, we run Algorithm 2 on the subtrees that contain e (there is at least one such tree and at most two). Let the output of the "even" instance be D^e, and the output of the "odd" instance be D^o. The output is $D^e \cup D^o$. Correctness follows from the fact that each (s_i, t_i) pair is completely contained in (at least) one of the subtrees by the assumption that $\text{dist}(s_i, t_i) \leq \ell$. Regarding the approximation ratio, recall that by Theorem 13, the capacity of each of D^e, D^o is no larger than twice the minimum multicut capacity, the capacity in the overall output is at most 4 times that of the minimal capacity multicut.

To obtain a feasible flow, we start by splitting the capacity of each edge $e \in D$ between the subtrees it is a member of, such that in each subtree there are at

---

[7] The ratio is $\frac{1}{4}$ when all capacities are even, and it tends to $\frac{1}{4}$ as $c_{\min} \to \infty$. For $c_{\min} = 1$ the approximation ratio is 0.

least $\lfloor \frac{c_e}{2} \rfloor$ capacity units. This is possible because each edge is a member in at most 2 subtrees. Now, given a query e to the LCA, we run Algorithm 2 on the trees e is a member of, obtaining flow values $f^e(e)$ for the "even" subtree and $f^o(e)$ for the "odd" subtree. The LCA outputs $f^e(e) + f^o(e)$.

The following is out main theorem for this section. Its proof appears in the full version of the paper.

Theorem 14. *Given a tree T with maximal degree d, integer edge capacities at least $c_{\min}$ and source-destination pairs with maximal distance at most $\ell > 0$, there are LCAs with $t(n) = d^{O(\ell)}$, $p(n) = d^{O(\ell)}$ and $em(n) = 0$, for 4-approximate multicut and $(\frac{1}{4} - \frac{1}{4c_{\min}})$-approximate IMCF. If all capacities are even, the approximation ratio to IMCF is $\frac{1}{4}$.*

4.2   Randomized LCA

We now turn to the randomized setting. Our randomized algorithm (see the full version for a more in-depth description and pseudocode) is similar: instead of an overlapping decomposition, we use a random one as follows. Let $H = \lceil \frac{\ell}{\epsilon} \rceil$. We pick an integer j uniformly at random from $[H]$, and remove all edges whose depth modulo H is $j - 1$. The result is a collection of subtrees of depth at most $H - 1$ each. Now, given an edge e, we run Algorithm 2 on the subtree that contains e and output the output of Algorithm 2 (with probability $1/H$, the edge queried, e, is not in any tree; in this case, e carries 0 flow).

Theorem 15. *Given a tree T with maximal degree d, integer edge capacities and vertex pairs with maximal distance at most $\ell > 0$, there is an LCA with $t(n) = d^{O(\ell/\epsilon)}$, $p(n) = d^{O(\ell/\epsilon)}$, and $em(n) = O(\log \ell/\epsilon)$, that achieves an approximation ratio of $(1/2 - \epsilon)$ to IMCF.*

5   Weighted Matchings

In this section we present a different kind of an LCA: a reduction. Specifically, we consider the task of computing a maximum weight matching (MWM), and show how to *locally* reduce it to maximum cardinality matching (MCM). Our construction, given any graph of maximal degree d and a t-time α-approximation LCA for MCM, yields an $O(td)$-time, $\frac{\alpha}{8}$-approximation LCA for MWM.

Formally, in MWM we are given a graph $G = (V, E)$ with a weight function $w : E \rightarrow \mathbb{N}$, and we need to output a set of disjoint edges of (approximately) maximum total *weight*. In MCM, the task is to find a set of disjoint edges of (approximately) the largest possible *cardinality*.

The main idea in our reduction is a variant of the well-known technique of *scaling* (e.g., [5,15,17]): partition the edges into classes of more-or-less uniform weight, run an MCM instance for each class, and somehow combine the MCM outputs. Motivated by local computation, however, we use a very crude combining rule that lends itself naturally to LCAs.

Specifically, the algorithm is as follows (the "global" algorithm is presented as Algorithm 3). Let $\gamma = 4$. Partition the edges by weight to sets E_i, such that $E_i = \{e : w(e) \in [\gamma^{i-1}, \gamma^i)\}$. For each i, find a maximum cardinality matching M_i on the graph $G_i = (V, E_i)$, using any MCM algorithm. Let $M = \cup_i M_i$. Given an edge e, our LCA for MWM returns "yes" iff e is a local maximum in M, i.e., iff (1) e is in M, and (2) for any edge e' in M which shares a node with e, $w(e') < w(e)$ (no ties can occur).

Algorithm 3. Reduction of MWM to MCM

Input : A graph $G = (V, E)$, with $w : E \to \mathbb{N}$, and $\gamma > 2$
Output: A matching M

Partition the edges into *classes*
$E_i = \{e : w(e) \in [\gamma^{i-1}, \gamma^i)\}$ for $i = 1, 2, \ldots$
In parallel, compute an unweighted matching M_i for each level i;
$M = \bigcup_i M_i$;
for *each edge $e \in M$* **do**
  if *e has a neighbor $e' \in M$, with* $\text{class}(e') < \text{class}(e)$ **then**
    $\llcorner$ Remove e from M;
Return M.

Our main result for this section is the following theorem, whose proof can be found in the full version of the paper.

Theorem 16. *Let $\mathcal{A}$ be a LCA for unweighted matching, requiring $t(n)$ time, $p(n)$ probes and $em(n)$ enduring memory, and producing an α-approximation to the maximum matching. Then given a graph $G = (V, E)$ with maximal degree d and arbitrary weights on the edges, there is a LCA that computes a $\alpha/8$-approximation to the maximum weighted matching, requiring $O(d \cdot t(n))$ time, $O(d \cdot p(n))$ probes and $O(d \cdot em(n))$ enduring memory.*

Acknowledgements. The authors would like to thank the anonymous reviewers for their useful feedback.

References

1. Alon, N., Rubinfeld, R., Vardi, S., Xie, N.: Space-efficient local computation algorithms. In: Proceedings of the 22nd ACM-SIAM Symposium on Discrete Algorithms (SODA), pp. 1132–1139 (2012)
2. Even, G., Medina, M., Ron, D.: Deterministic stateless centralized local algorithms for bounded degree graphs. In: Schulz, A.S., Wagner, D. (eds.) ESA 2014. LNCS, vol. 8737, pp. 394–405. Springer, Heidelberg (2014)
3. Garg, N., Vazirani, V.V., Yannakakis, M.: Primal-dual approximation algorithms for integral flow and multicut in trees. Algorithmica **18**(1), 3–20 (1997)
4. Linial, N.: Locality in distributed graph algorithms. SIAM J. Comput. **21**(1), 193–201 (1992)
5. Lotker, Z., Patt-Shamir, B., Rosén, A.: Distributed approximate matching. SIAM J. Comput. **39**(2), 445–460 (2009)

6. Mansour, Y., Rubinstein, A., Vardi, S., Xie, N.: Converting online algorithms to local computation algorithms. In: Czumaj, A., Mehlhorn, K., Pitts, A., Wattenhofer, R. (eds.) ICALP 2012, Part I. LNCS, vol. 7391, pp. 653–664. Springer, Heidelberg (2012)
7. Mansour, Y., Vardi, S.: A local computation approximation scheme to maximum matching. In: Raghavendra, P., Raskhodnikova, S., Jansen, K., Rolim, J.D.P. (eds.) RANDOM 2013 and APPROX 2013. LNCS, vol. 8096, pp. 260–273. Springer, Heidelberg (2013)
8. Nešetřil, J., Milková, E., Nešetřilová, H.: Otakar Borůvka on minimum spanning tree problem: Translation of both the 1926 papers, comments, history. Discrete Math. **233**(1), 3–36 (2001)
9. Nguyen, H.N., Onak, K.: Constant-time approximation algorithms via local improvements. In: Proceedings of the 49th Annual IEEE Symposium on Foundations of Computer Science (FOCS), pp. 327–336 (2008)
10. Panconesi, A., Rizzi, R.: Some simple distributed algorithms for sparse networks. Distrib. Comput. **14**(2), 97–100 (2001)
11. Reingold, O., Vardi, S.: New techniques and tighter bounds for local computation algorithms (2015). Under submission
12. Rubinfeld, R., Tamir, G., Vardi, S., Xie, N.: Fast local computation algorithms. In: Proceedings of the 2nd Symposium on Innovations in Computer Science (ICS), pp. 223–238 (2011)
13. Suomela, J.: Survey of local algorithms. ACM Comput. Surv. **45**(2), 24 (2013)
14. Tarjan, R.E.: Data Structures and Network Algorithms. Society for Industrial and Applied Mathematics, Philadelphia (1983)
15. Uehara, R., Chen, Z.-Z.: Parallel approximation algorithms for maximum weighted matching in general graphs. In: Watanabe, O., Hagiya, M., Ito, T., Leeuwen, J., Mosses, P.D. (eds.) TCS 2000. LNCS, vol. 1872, pp. 84–98. Springer, Heidelberg (2000)
16. Vazirani, V.V.: Approximation Algorithms. Springer, Heidelberg (2001)
17. Wattenhofer, M., Wattenhofer, R.: Distributed weighted matching. In: Guerraoui, R. (ed.) DISC 2004. LNCS, vol. 3274, pp. 335–348. Springer, Heidelberg (2004)
18. Yoshida, Y., Yamamoto, M., Ito, H.: Improved constant-time approximation algorithms for maximum matchings and other optimization problems. SIAM J. Comput. **41**(4), 1074–1093 (2012)

An $O(\log \mathrm{OPT})$-Approximation
for Covering/Packing Minor Models of θ_r

Dimitris Chatzidimitriou[1]([✉]), Jean-Florent Raymond[2,3],
Ignasi Sau[2], and Dimitrios M. Thilikos[1,2]

[1] Department of Mathematics,
National and Kapodistrian University of Athens, Athens, Greece
hatzisdimitris@gmail.com, sedthilk@thilikos.info
[2] AlGCo Project-Team, CNRS, LIRMM, Montpellier, France
jean-florent.raymond@mimuw.edu.pl, sau@lirmm.fr
[3] Computer Science Institute, University of Warsaw, Warsaw, Poland

Abstract. Let $\mathcal{C}_H$ be the class of graphs containing some fixed graph H as a minor. We define $\mathbf{c}_H^{\mathsf{v}}(G)$ (resp. $\mathbf{c}_H^{\mathsf{e}}(G)$) as the minimun number of vertices (resp. edges) whose removal from G produces a graph without any subgraph isomorphic to a graph in $\mathcal{C}_H$. Also $\mathbf{p}_H^{\mathsf{v}}(G)$ (resp. $\mathbf{p}_H^{\mathsf{e}}(G)$) is the the maximum number of vertex- (resp. edge-) disjoint subgraphs of G that are isomorphic to some graph in $\mathcal{C}_H$. We denote by θ_r the graph with two vertices and r parallel edges between them. When $H = \theta_r$, the parameters $\mathbf{c}_H^{\mathsf{v/e}}$ and $\mathbf{p}_H^{\mathsf{v/e}}$ are NP-complete to compute (for sufficiently large r). In this paper we prove a series of combinatorial and algorithmic lemmata that imply that if $\mathbf{p}_{\theta_r}^{\mathsf{v/e}}(G) \leq k$, then $\mathbf{c}_{\theta_r}^{\mathsf{v/e}}(G) = O(k \log k)$. This means that for every r, the class $\mathcal{C}_{\theta_r}$ has the vertex/edge Erdős-Pósa property. Using the combinatorial ideas from our proofs we introduce a unified approach for the design of an $O(\log \mathrm{OPT})$-approximation algorithm for $\mathbf{c}_{\theta_r}^{\mathsf{v}}$, $\mathbf{p}_{\theta_r}^{\mathsf{v}}$, $\mathbf{c}_{\theta_r}^{\mathsf{e}}$ and $\mathbf{p}_{\theta_r}^{\mathsf{e}}$ that runs in $O(n \cdot \log(n) \cdot m)$ steps.

Keywords: Erdős-Pósa properties · Minors · Graph packing · Covering

1 Introduction

All graphs in this paper are undirected, do not have loops but they may contain multiple edges. We denote by θ_r the graph containing two vertices x and y connected by r parallel edges. Given a graph class $\mathcal{C}$ and a graph G, we call

The results presented in this extended abstract have appeared, among others, in Arxiv as [8] and [7]. The work of the first and the last author is co-financed by the European Union (European Social Fund - ESF) and Greek national funds through the Operational Program "Education and Lifelong Learning" of the National Strategic Reference Framework (NSRF) - Research Funding Program: Thales. Investing in knowledge society through the European Social Fund. The work of the third author is co-financed by the (Polish) National Science Centre grant PRELUDIUM DEC-2013/11/N/ST6/02706.

L. Sanità and M. Skutella (Eds.): WAOA 2015, LNCS 9499, pp. 122–132, 2015.
DOI: 10.1007/978-3-319-28684-6_11

$\mathcal{C}$-*subgraph* of G any subgraph of G that is isomorphic to some graph in $\mathcal{C}$. In this paper, when giving the running time of an algorithm with input some graph G, we agree that $n = |V(G)|$ and $m = |E(G)|$.

Coverings and Packings. Paul Erdős and Lajos Pósa, proved in 1965 [12] that there is a function $f \colon \mathbb{N} \to \mathbb{N}$ such that for each positive integer k, every graph either contains k vertex-disjoint cycles or it contains $f(k)$ vertices that intersect every cycle in G. Moreover, they proved that the "gap" of this min-max relation is $f(k) = O(k \cdot \log k)$. This result initiated an interesting line of research on the duality between coverings and packings of combinatorial objects. To formulate this duality, given a class $\mathcal{C}$ of connected graphs, we define by $\mathbf{c}_{\mathcal{C}}^{\mathsf{v}}(G)$ (resp. $\mathbf{c}_{\mathcal{C}}^{\mathsf{e}}(G)$) the minimun cardinality of a set S of vertices (resp. edges) such that each $\mathcal{C}$-subgraph of G contains some element of S. Also, we define $\mathbf{p}_{\mathcal{C}}^{\mathsf{v}}(G)$ (resp. $\mathbf{p}_{\mathcal{C}}^{\mathsf{e}}(G)$) as the maximum number of vetex- (resp. edge-) disjoint $\mathcal{C}$-subgraphs of G.

We say that $\mathcal{C}$ has the *vertex Erdős-Pósa property* (resp. the *edge Erdős-Pósa property*) if there is a function $f \colon \mathbb{N} \to \mathbb{N}$, called *gap function*, such that for every graph G, $\mathbf{c}_{\mathcal{C}}^{\mathsf{v}}(G) \leq f(\mathbf{p}_{\mathcal{C}}^{\mathsf{v}}(G))$ (resp. $\mathbf{c}_{\mathcal{C}}^{\mathsf{e}}(G) \leq f(\mathbf{p}_{\mathcal{C}}^{\mathsf{e}}(G))$). Using this terminology, the original result of Erdős and Pósa says that the set of all cycles has the vertex Erdős-Pósa property with gap $O(k \cdot \log k)$. The general question in this branch of Graph Theory is to detect instantiations of $\mathcal{G}$ which have the vertex/edge Erdős-Pósa property (in short, v/e-*EP-property*) and, when this is the case, minimize the gap function f. Several theorems of this type have been proved concerning different instantiations of $\mathcal{G}$ such as odd cycles [22,28], long cycles [3], and graphs containing cliques as minors [11] (see also [17,20,30] for results on more general combinatorial structures).

A general class that is known to have the v-EP-property is the class $\mathcal{C}_H$ of the graphs that contain some fixed planar graph H as a minor[1]. This fact was proven by Robertson and Seymour in [31] and the best known general gap is $f(k) = O(k \cdot \log^{O(1)} k)$ due to the results of [9] (see also [14,15] for better gaps for particular instantiations of H). Moreover, the planarity of H appears to be the right dichotomy, as for non-planar H, $\mathcal{C}_H$ does not have the v-EP-property. Besides the near-optimality of the general upper bound of [9], it is open whether the lower bound $\Omega(k \cdot \log k)$ can be matched for the general gap function, while this is indeed the case when $H = \theta_r$ [14].

The question about classes that have the e-EP-property has also attracted some attention (see [3]). According to [10, Exercice 23 of Chapter7], the original proof of Erdős and Pósa implies that cycles have the e-EP-property with gap $O(k \cdot \log k)$. Moreover, as proved in [29], the class $\mathcal{C}_{\theta_r}$ has the e-EP-property with the (non-optimal) gap $f(k) = O(k^2 \cdot \log^{O(1)} k)$. Interestingly, not much more is known on the graphs H for which $\mathcal{G}_H$ has the e-EP-property and it is tempting to conjecture that the planarity of H provides again the right dichotomy. Other graph classes that are known to have the e-EP-property are rooted cycles (here the cycles to be covered and packed are required to intersect some particular set

---

[1] A graph H is a *minor* of a graph G if it can be obtained from some subgraph of G by contracting edges.

of terminals of G) [27] and odd cycles for the case where G is a 4-edge-connected graph [21], or a planar graph [24], or a graph embeddable in an orientable surface [22].

Approximation Algorithms. The four graph parameters defined above are already quite general when $\mathcal{C} := \mathcal{C}_H$. For simplicity, form now on, we use the notations $\mathbf{c}_H^v$, $\mathbf{c}_H^e$, $\mathbf{p}_H^v$, and $\mathbf{p}_H^e$ instead of $\mathbf{c}_{\mathcal{C}_H}^v$, $\mathbf{c}_{\mathcal{C}_H}^e$, $\mathbf{p}_{\mathcal{C}_H}^v$, and $\mathbf{p}_{\mathcal{C}_H}^e$. From the algorithmic point of view, the computation of $\mathbf{p}_H^{v/e}$ corresponds to the general family of graph packing problems while the computation of $\mathbf{c}_H^{v/e}$ belongs to the general family of graph modification problems where the modification operation is the removal of vertices/edges. Interestingly, particular instantiations of $H = \theta_r$ generate known, well studied, NP-hard problems. For instance, asking whether $\mathbf{c}_{\theta_r}^v \leq k$ generates VERTEX COVER for $r = 1$, FEEDBACK VERTEX SET for $r = 2$, and DIAMOND HITTING SET for $r = 3$ [13,16]. Moreover, asking whether $\mathbf{p}_{\theta_2}^x(G) \geq k$ corresponds to VERTEX CYCLE PACKING [6,23] and EDGE CYCLE PACKING [1,25] when x = v and x = e respectively. Finally, asking whether $|E(G)| - \mathbf{c}_{\theta_3}^e(G) \leq k$ corresponds to the MAXIMUM CACTUS SUBGRAPH[2]. All parameters keep being NP-complete because the aforementioned base cases can be reduced to the general one by replacing each edge by one of multiplicity $r-1$.

From the approximation point of view, it was proven in [16], that when H is a planar graph, there is a randomized polynomial $O(1)$-approximation algorithm for $\mathbf{c}_H^v$. For the cases of $\mathbf{c}_{\theta_r}^v$ and $\mathbf{p}_{\theta_r}^v$, $O(\log n)$-approximations are known for every $r \geq 1$ because of [19] (see also [32]). Moreover, $\mathbf{c}_{\theta_3}^v$ admits a deterministic 9-approximation [13]. About $\mathbf{p}_{\theta_r}^e(G)$ it is known, from [26], that there is a polynomial $O(\sqrt{\log n})$-approximation algorithm for the case where $r = 2$. Notice also that, it is trivial to compute $\mathbf{c}_{\theta_2}^e(G)$ in polynomial time. However, to our knowledge, nothing is known about the computation of $\mathbf{c}_{\theta_r}^e(G)$ for $r \geq 3$.

Our Results. We introduce a unified approach for the study of the combinatorial interconnections and the approximability of the parameters $\mathbf{c}_{\theta_r}^v$, $\mathbf{c}_{\theta_r}^e$, $\mathbf{p}_{\theta_r}^v$, and $\mathbf{p}_{\theta_r}^e$. Our main combinatorial result is the following.

Theorem 1. *For every $r \in \mathbb{N}_{\geq 2}$ and every x $\in \{v, e\}$ the graph class $\mathcal{C}_{\theta_r}$ has the x-EP-property with (optimal) gap function $f(k) = O(k \cdot \log k)$.*

Our proof is unified and treats simultaneously the covering and the packing parameters for both the vertex and the edge cases. This verifies the best, so far, combinatorial bound for the case where x = v [14] and optimally improves the previous bound in [29] for the case where x = e. Based on the proof of Theorem 1, we prove the following algorithmic result.

---

[2] The MAXIMUM CACTUS SUBGRAPH problem asks, given a graph G and an integer k, whether G contains a subgraph of k edges where no two cycles share an edge. It is easy to reduce to this problem the VERTEX CYCLE PACKING problem on cubic graphs which, in turn, can be proved to be NP-complete using a simple variant of the NP-completeness proof of the EXACT COVER BY 2-SETS [18]).

Theorem 2. *For every $r \in \mathbb{N}_{\geq 2}$ and every $\mathsf{x} \in \{\mathsf{v}, \mathsf{e}\}$ there exists an $O(n \cdot \log(n) \cdot m)$-step algorithm that, given a graph G, outputs a $O(\log \mathrm{OPT})$-approximation for $\mathbf{c}^{\mathsf{x}}_{\theta_r}$ and $\mathbf{p}^{\mathsf{x}}_{\theta_r}$.*

Theorem 2 improves the results in [19] for the cases of $\mathbf{c}^{\mathsf{v}}_{\theta_r}$ and $\mathbf{p}^{\mathsf{v}}_{\theta_r}$ and, to our knowledge, is the first approximation algorithm for $\mathbf{c}^{\mathsf{e}}_{\theta_r}$ and $\mathbf{p}^{\mathsf{e}}_{\theta_r}$ for $r \geq 3$.

Our Techniques. Our proofs are based on two lemmata: a combinatorial (Lemma 1) and an algorithmic one (Lemma 2). The combinatorial Lemma 1 asserts that given a graph G, one of the following holds:

1. either G contains k disjoint $\mathcal{C}_{\theta_r}$-subgraphs; or
2. it contains a $\mathcal{C}_{\theta_r}$-subgraph J of at most $O(\log k)$ edges; or
3. it contains a subset X on at least $f(r)$ vertices that induces a "tree-like" subgraph that is separated by $2r - 2$ edges from the rest of the graph (we call these subgraphs *edge-protrusions*, in analogy to the notion of vertex protrusions that where introduced in [4,5]).

Moreover, each of the three above possible outcomes can be produced in $|E(G)|$ steps.

Intuitively, the above result says that we can reduce the inputs of the problem of computing $\mathbf{c}^{\mathsf{x}}_{\theta_r}$ or $\mathbf{p}^{\mathsf{x}}_{\theta_r}$ to "loosely connected" graphs, i.e., graphs where the removal of $O(r)$ edges cannot break them into two parts of "big" size (as a function of r). In such graphs, we can restrict our attention to packings or coverings of $\mathcal{C}_{\theta_r}$-subgraphs that have at most $O(\log(\mathbf{p}^{\mathsf{x}}_{\theta_r}(G)))$ edges. This easily yields that $\mathcal{C}_{\theta_r}$ has the x-*EP-property* for every $\mathsf{x} \in \{\mathsf{v}, \mathsf{e}\}$ with gap $O(k \cdot \log k)$. Our main algorithm is doing the following for each $k \leq |V(G)|$. If the first case of the above combinatorial result applies on G, we can safely output a packing of k $\mathcal{C}_{\theta_r}$-subgraphs in G. In the second case, we make some progress as we may remove the vertices/edges of J from G and then set $k := k - 1$. In the third case, we again progress as we can prove that if X is big enough then we can replace G, in a linear number of steps, by another smaller graph G' where both packing and covering numbers stay the same.

Notice that the second step reduces the packing number of the current graph by 1 to the price of reducing the covering number by $O(\log k)$ and this is the main argument that supports the claimed $O(\log \mathrm{OPT})$-approximation algorithm. Finally, the third step is supported on the algorithmic Lemma 2 whose proof is based on a dynamic programming encoding of partial packings and coverings that is designed especially for the corresponding tree-like graphs.

2 Definitions and Preliminaries

As graphs in this paper may have multi-edges, they are represented by a pair $G = (V, E)$ where V is its vertex set, denoted by $V(G)$ and E is its edge multi-set, denoted by $E(G)$.

Let $\mathsf{x} \in \{\mathsf{v}, \mathsf{e}\}$ where v is interpreted as "vertex" and e will be interpreted as "edge". Given a graph G, we denote by $A_{\mathsf{x}}(G)$ the set of vertices or edges of G depending on whether $\mathsf{x} = \mathsf{v}$ or $\mathsf{x} = \mathsf{e}$. We set $n(G) = |V(G)|$ and $m(G) = |E(G)|$.

Topological Minors. Two paths of a graph P are *internally vertex disjoint* if the one does not contain any of the internal vertices of the other. We say that H is a *topological minor* of G if there exits some collection $\mathcal{P}$ of pairwise internally vertex disjoint paths in G and a injection $\phi : V(H) \to V(G)$ such that

- no path in $\mathcal{P}$ has an internal vertex that belong to some other path in $\mathcal{P}$;
- $\phi(V(H))$ is the set of endpoints of the paths in $\mathcal{P}$; and
- for every two distinct $v, u \in V(H)$, $\{v, u\}$ is an edge of H of multiplicity l if and only if there are l paths in $\mathcal{P}$ between $\phi(v)$ and $\phi(u)$.

Packings and Coverings. If G is a graph and $\mathcal{H}$ is a (finite) class of connected graphs, an $\mathcal{H}$-*model of G* is a subgraph M of G that is a subdivision of a graph, denoted by $\hat{M}$, that is isomorphic to a member of $\mathcal{H}$. Clearly, the vertices of $\hat{M}$ are vertices of G and its edges correspond to paths in G between their endpoints such that internal vertices of a path do not appear in any other path. We refer to the vertices of $\hat{M}$ in G as the *branch vertices* of the $\mathcal{H}$-model M, whereas internal vertices of the paths between branch vertices will be called *subdivision vertices* of M. A graph which contains no $\mathcal{H}$-model is said to be $\mathcal{H}$-free. Notice that G has a $\mathcal{H}$-model iff G contains a graph of $\mathcal{H}$ as topological minor. Given a $\mathsf{x} \in \{\mathsf{v}, \mathsf{e}\}$, an x-$\mathcal{H}$-*packing* of a graph G is a collection of pairwise x-disjoint $\mathcal{H}$-models of G. We define $\mathbf{P}_{\mathsf{x}, \mathcal{H}}^{\geq k}(G)$ as the set of all x-$\mathcal{H}$-packings of G of size at least k. An x-$\mathcal{H}$-*covering* of a graph G is a set $C \subseteq A_{\mathsf{x}}(G)$ such that $G \setminus C$ does not contain any $\mathcal{H}$-model. We define $\mathbf{C}_{\mathsf{x}, \mathcal{H}}^{\leq k}(G)$ as the set of all x-H-coverings of G of size at most k. We define $\mathsf{x}\text{-cover}_{\mathcal{H}}(G) = \min\{k, \ \mathbf{C}_{\mathsf{x}, \mathcal{H}}^{\leq k}(G) \neq \emptyset\}$ and $\mathsf{x}\text{-pack}_{\mathcal{H}}(G) = \max\{k, \ \mathbf{P}_{\mathsf{x}, \mathcal{H}}^{\geq k}(G) \neq \emptyset\}$. Given a graph H, we define by $\mathsf{ex}(H)$ the set of all topologically-minor minimal graphs that contain H as a minor. Notice that the size of $\mathsf{ex}(H)$ is bounded by some function of $m(H)$. It is easy to verify that H is a minor of G if and only if it contains a member of $\mathsf{ex}(H)$ as a topological minor. An H-*minor model of G* is every minimal subgraph of G that contains a member of $\mathsf{ex}(H)$ as a topological minor. Notice that

$$\mathbf{c}_H^{\mathsf{x}}(G) = \mathsf{x}\text{-cover}_{\mathsf{ex}(H)}(G) \quad \text{and} \quad \mathbf{p}_H^{\mathsf{x}}(G) = \mathsf{x}\text{-pack}_{\mathsf{ex}(H)}(G). \tag{1}$$

The above equalities gives us the right, in the rest of this paper, to deal with our results and describe proofs in terms of $\mathsf{x}\text{-cover}_{\mathsf{ex}(H)}$ and $\mathsf{x}\text{-pack}_{\mathsf{ex}(H)}$ instead of $\mathbf{c}_H^{\mathsf{x}}$ and $\mathbf{p}_H^{\mathsf{x}}$.

Edge-protrusions and Tree Partitions. A *rooted tree* is a pair (T, s) such that $s \in V(T)$. A *rooted tree partition* of a graph G is a triple $\mathcal{D} = (\mathcal{X}, T, s)$ where (T, s) is a rooted tree and $\mathcal{X} = \{X_i, \ i \in V(T)\}$ is a partition of $V(G)$ where either $n(T) = 1$ or for every $\{x, y\} \in E(G)$, there exists an edge $\{i, j\} \in E(T)$ such that $\{x, y\} \subseteq X_i \cup X_j$. Given an edge $f = \{i, j\} \in E(T)$, we define E_f as the set of edges with one endpoint in X_i and the other in X_j. Notice that all edges in E_f are non-loop edges. The *width* of $\mathcal{D}$ is defined as $\max\{|X_i|, \ i \in V(T)\} \cup \{|E_f|, \ f \in E(T)\}$.

Let G be a graph and let $Y \subseteq V(G)$. We say that Y is a *t-edge-protrusion* of G with *extension* w if the graph $G[Y \cup N_G(Y)]$ has a rooted tree partition $\mathcal{D} = (\mathcal{X}, T, s)$ of width at most t and such that $N_G(Y) = X_s$ and $n(T) \geq w$.

2.1 Two Basic Lemmata

Let $x_1, \ldots, x_l \in \mathbb{N}$ and $\psi, \chi : \mathbb{N} \to \mathbb{N}$. We say that $\chi(n) = O_{x_1,\ldots,x_l}(\psi(n))$ if there exists a computable function $\phi : \mathbb{N}^l \to \mathbb{N}$ such that $\chi(n) = O(\phi(x_1, \ldots, x_l) \cdot \psi(n))$.

Our results are based on the following two lemmata (the fool proofs are in [7] and [8] respectively).

Lemma 1 (Combinatorial Lemma – Theorem *3* in *[7]*). *There is an algorithm that, with input three positive integers r, w, z and an n-vertex graph W, outputs one of the following:*

- *a θ_r-model of W of at most z edges;*
- *a connected $(2r - 2)$-edge-protrusion Y of W with extension $> w$; or*
- *an H-minor model of W for some graph H where $\delta(H) \geq \frac{1}{r-1} 2^{\frac{z-5r}{4r(2w+1)}}$,*

in $O_r(m)$ steps.

Lemma 2 (Algorithmic Lemma – Lemma *1* in *[8]*). *Let $\mathrm{x} \in \{\mathrm{v}, \mathrm{e}\}$, $h, t \in \mathbb{N}$. There exists a function $f_1 : \mathbb{N}^2 \to \mathbb{N}$ and an algorithm that receives as input a collection $\mathcal{H}$ of connected graphs that, in total, have h edges, a graph W, and a t-edge protrusion Y of W with extension at least $f_1(h, t)$ and outputs*

- *either an $\mathcal{H}$-model of W of at most $f_1(h, t)$ edges;*
- *or a graph W' where $\mathrm{x\text{-}cover}_{\mathcal{H}}(W) = \mathrm{x\text{-}cover}_{\mathcal{H}}(W')$, $\mathrm{x\text{-}pack}_{\mathcal{H}}(W) = \mathrm{x\text{-}pack}_{\mathcal{H}}(W')$, and $n(W') < n(W)$,*

in $O_h(n(Y))$ steps.

3 Algorithms for Theorems 1 and 2

For the purposes of this section we define $\Theta_r = \mathrm{ex}(\theta_r)$. In this section we give the proofs of Theorems 1 and 2. Because of (1), all proofs will use the terms $\mathrm{x\text{-}cover}_{\Theta_r}$ and $\mathrm{x\text{-}pack}_{\Theta_r}$ instead of $\mathbf{c}^{\mathrm{x}}_{\theta_r}(G)$ and $\mathbf{p}^{\mathrm{x}}_{\theta_r}(G)$. For every $i, j \in \mathbb{N}$, we use $[\![i, j]\!]$ to denote the interval $\{t \in \mathbb{N}, \ i \leq t \leq j\}$.

3.1 Erdő-Pósa propery for θ_r

We need the following result.

Proposition 1 (Theorem 12 of *[2]*). *Given $k, r \in \mathbb{N}_{\geq 1}$ and an input graph G such that $\delta(G) \geq k(r + 1) - 1$, a partition $(V_1, \ldots, V_k)$ of $V(G)$ satisfying $\forall i \in [\![1, k]\!]$, $\delta(G[V_i]) \geq r$ can be found in $O(n^c)$ steps for some $c \in \mathbb{N}$.*

Combining Lemmas 1 and 2, we get the following result.

Lemma 3. *There is an algorithm that, with input* $x \in \{v, e\}$, $r \in \mathbb{N}_{\geq 2}$, $k \in \mathbb{N}$ *and an n-vertex graph W, outputs one of the following:*

- *a Θ_r-model of W of at most $O_r(\log k)$ edges;*
- *a graph W' where $n(W') < n(W)$, $x\text{-cover}_{\mathcal{H}}(W') = x\text{-cover}_{\mathcal{H}}(W)$, and $x\text{-pack}_{\mathcal{H}}(W') = x\text{-pack}_{\mathcal{H}}(W)$;*
- *an H-minor model in W for some graph H where $\delta(H) \geq k(r+1)$,*

in $O_r(m)$ steps.

Proof. We set $t = 2r - 2$, $w = f_1(h, t)$, $z = 2r(w-1)\log(k(r+1)(r-1)) + 5r$ and $h = m(\Theta_r)$. Remark that $z = O_r(\log k)$ and $h, t, w = O_r(1)$. Also observe that our choice of z ensures that $2^{\frac{z-5r}{2r(w-1)}}/(r-1) = k(r+1)$.

By applying the algorithm of Lemma 1 to r, w, z and W, we obtain in $O_r(m(W))$-time:

First case: either a Θ_r-model in W of at most z edges;
Second case: or a $(2r-2)$-edge-protrusion Y of W with extension $> w$;
Third case: or an H-minor model M in W for some graph H where $\delta(H) \geq k(r+1)$.

- In the first case, we return the obtained Θ_r-model.
- In the second case, by applying the algorithm of Lemma 2 on Y, we get in $O(n(W))$-time either a Θ_r-model of W on at most $w = O_r(1)$ vertices, or a graph W' where, for $x \in \{v, e\}$, $x\text{-cover}_{\mathcal{H}}(W') = x\text{-cover}_{\mathcal{H}}(W)$, $x\text{-pack}_{\mathcal{H}}(W') = x\text{-pack}_{\mathcal{H}}(W)$ and $n(W') < n(W)$.
- In the third case, we return the model M.

In each of the above cases, we get after $O(m)$ steps either a model of a graph with minimum degree more than $k(r+1)$, or a Θ_r-model in W of at most z edges, or an equivalent graph of smaller size.   □

It might not be clear yet to what purpose the the model of a graph of degree more than $k(r+1)$ output by the algorithm of Lemma 3 can be used. An answer is given by the following lemmata, which state that such a graph contains a packing of at least k models of Θ_r. These lemmata will be used below for the design of the approximation algorithms.

Lemma 4. *There is an algorithm that, given $k, r \in \mathbb{N}_{\geq 1}$ and a graph G with $\delta(G) \geq kr$, returns a member of $\mathbf{P}_{e,\Theta_r}^{\geq k}(G)$ in G in $O(m)$ steps.*

Proof. Starting from any vertex u, we grow a maximal path P in G by iteratively adding to P a vertex that is adjacent to the previously added vertices but does not belong to P. Since $\delta(G) \geq kr$, any such path will have length at least $kr + 1$. At the end, all the neighbors of the last vertex v of P belong to P (otherwise P could be extended). Since v has degree at least kr, v has at least kr neighbors in P. Let $w_0, \ldots, w_{kr-1}$ be an enumeration of the kr first neighbors of v in the order given by P, starting from u. For every $i \in [\![0, k-1]\!]$, let S_i be the subgraph of G induced by v and the subpath of P starting at w_{ir} and ending at $w_{(i+1)r-1}$. Observe that for every $i \in [\![0, k-1]\!]$, S_i contains a Θ_r-model and that the intersection of every pair of graphs from $\{S_i\}_{i \in [\![0,k-1]\!]}$ is $\{v\}$. Hence P contains a member of $\mathbf{P}_{e,\Theta_r}^{\geq k}(G)$, as desired. Every edge of G is crossed at most once in this algorithm, yielding to a running time of $O(m)$ steps.   □

Corollary 1. *There is an algorithm that, given $r \in \mathbb{N}_{\geq 1}$ and a graph G with $\delta(G) \geq r$, returns a Θ_r-model in G in $O(m)$-steps.*

An analogue of Lemma 4 for vertex-disjoint packings can be proved using Proposition 1.

Lemma 5. *There is an algorithm that, given $k, r \in \mathbb{N}_{\geq 1}$ and a graph G with $\delta(G) \geq k(r+1) - 1$, outputs a member of $\mathbf{P}_{v,\Theta_r}^{\geq k}(G)$ in $O(n^c + m)$ steps, where c is the constant of Proposition 1.*

Proof. After applying the algorithm of Proposition 1 on G to obtain in $O(n^c)$-time k graphs $G[V_1], \ldots, G[V_k]$, we extract a Θ_r-model from each of them using Corollary 1.   □

Theorem 1 follows immediately from the following more general result.

Theorem 3. *There is a function $f_2 \colon \mathbb{N} \to \mathbb{N}$ and an algorithm that, with input $\mathsf{x} \in \{\mathsf{v}, \mathsf{e}\}$, $r \in \mathbb{N}_{\geq 2}$, $k \in \mathbb{N}$ and an n-vertex graph W, outputs either a x-Θ_r-packing of W of size k or an x-Θ_r-covering of W of size $f_2(r) \cdot k \cdot \log k$. Moreover, this algorithm runs in $O(n \cdot m)$ steps if $\mathsf{x} = \mathsf{e}$ and in $O(n^c + n \cdot m)$ steps if $\mathsf{x} = \mathsf{v}$, where c is the constant Proposition 1.*

Proof. Let f_2 be a function such that each Θ_r-model output by the algorithm of Lemma 3 has size at most $f_2(r) \cdot \log k$. We consider the following procedure.

1. $G := W$; $P := \emptyset$;
2. apply the algorithm of Lemma 3 on (x, r, k, G):
   Progress: if the output is a Θ_r-model M, let $G := G \setminus A_{\mathsf{x}}(M)$ and $P = P \cup \{M\}$;
   Win: if the output is a H-minor model M in W for some graph H where $\delta(H) \geq k(r+1)$, apply the algorithm of Lemma 4 (if $\mathsf{x} = \mathsf{e}$) or the one of Lemma 5 (if $\mathsf{x} = \mathsf{v}$) to H to obtain a member of $\mathbf{P}_{\mathsf{x},\Theta_r}^{\geq k}(H)$. Using M, translate this packing into a member of $\mathbf{P}_{\mathsf{x},\Theta_r}^{\geq k}(W)$ and return this new packing;
   Reduce: otherwise, the output is a graph G' then let $G := G'$.
3. if $|P| = k$ then return P which is a member of $\mathbf{P}_{\mathsf{x},\Theta_r}^{\geq k}(W)$;
4. if $n(W) = 0$ then return P which, in this case, is a member of $\mathbf{C}_{\mathsf{x},\Theta_r}^{\leq f_2(r)k \log k}(W)$;
5. otherwise, loop to line 2.

This algorithm clearly returns the desired result. Furthermore, the loop is executed at most $n(W)$ times and each call of the algorithm of Lemma 3 takes $O(m(W))$ steps. When the algorithm reaches the "Win" case (which happen at most once), the calls to the algorithm of Lemma 4 (if $x = e$) or the one of Lemma 5 (if $x = v$) respectively take $O(m(H))$ and $O((n(H))^c)$ steps. Therefore in total, this algorithm terminates in $O(n \cdot m)$ steps if $x = e$ and in $O(n^c + n \cdot m)$ steps if $x = v$. $\qquad \square$

3.2 An Approximation Algorithm for Theorem 2

Observe that if the algorithm of Theorem 3 reaches the "Win" case, then the input graph is known to contains a x-Θ_r-packing of size at least k. As a consequence, if we are only interested in the existence of a packing or covering, the call to the algorithm of Lemmas 4 or 5 is not necessary.

Corollary 2. *There is an algorithm that, with input $x \in \{v, e\}$, $r \in \mathbb{N}_{\geq 2}$, $k \in \mathbb{N}$ and an graph W, outputs 0 if W has a x-Θ_r-packing of G of size k or 1 if W has a x-Θ_r-covering of G of size $f_2(r) \cdot k \cdot \log k$. Furthermore this algorithm runs in $O(n \cdot m)$ steps.*

We are now ready to prove Theorem 2.

Proof. (Proof of Theorem 2). Let us call A the algorithm of Corollary 2. Let $k_0 \in [\![1, n(W)]\!]$ be an integer such that $A((x, r, k_0, W) = 1$ and $A(x, r, k_0 - 1, W) = 0$. Our purpose is to prove that the value $k_0 \log k_0$ is a $O(\log OPT)$-approximation of $\mathbf{p}(W)$.

First, notice that for every $k > $x-pack$_{\Theta_r}(W)$, the value returned by $A(x, r, k, W)$ is 1. Symmetrically it holds that, for every k such that $k \log k < $x-cover$_{\Theta_r}(W)$, the value of $A(x, r, k, W)$ is 0. Therefore, the value k_0 is such that: $k_0 - 1 \leq $x-pack$_{\Theta_r}(W)$ and x-cover$_{\Theta_r}(W) \leq k_0 \log k_0$.

As every minimal covering must contain at least one vertex or edge (depending whether $x = v$ or $x = e$) of each model of a maximal packing x-pack$_{\Theta_r}(W) \leq$ x-cover$_{\Theta_r}(W)$, we have the two following equations:

$$\text{x-pack}_{\Theta_r}(W) \leq k_0 \log k_0 \leq (\text{x-pack}_{\Theta_r}(W) + 1) \log(\text{x-pack}_{\Theta_r}(W) + 1) \qquad (2)$$

$$\text{x-cover}_{\Theta_r}(W) \leq k_0 \log k_0 \leq (\text{x-cover}_{\Theta_r}(W) + 1) \log(\text{x-cover}_{\Theta_r}(W) + 1) \qquad (3)$$

Dividing (2) by x-pack$_{\Theta_r}(W)$ and (3) by cover$_{\Theta_r}(W)$, we have that: $1 \leq \frac{k_0 \log k_0}{\text{x-pack}_{\Theta_r}(W)} \leq \log(\text{x-pack}_{\Theta_r}(W) + 1) + \frac{\log \text{x-pack}_{\Theta_r}(W)}{\text{x-pack}_{\Theta_r}(W)} = O(\log(\text{x-pack}_{\Theta_r}(W)))$. Moreover, $1 \leq \frac{k_0 \log k_0}{\text{x-cover}_{\Theta_r}(W)} \leq \log(\text{x-cover}_{\Theta_r}(W) + 1) + \frac{\log \text{x-cover}_{\Theta_r}(W)}{\text{x-cover}_{\Theta_r}(W)} = O(\log(\text{x-cover}_{\Theta_r}(W)))$. Therefore the value $k_0 \log k_0$ is both a $O(\log OPT)$-approximation of x-pack$_{\Theta_r}(W)$ and cover$_{\Theta_r}(W)$. The value k_0 can be found by a binary search in the interval $[\![1, n]\!]$, in $O(\log n)$ calls to Algorithm A. Hence our approximation runs in $O(n \cdot \log(n) \cdot m)$ steps. $\qquad \square$

4   Discussion

Notice that Lemma 2 holds for every finite collection $\mathcal{H}$ of connected graphs. This means that the approximation algorithms of this paper could be extended for x-pack$_{\text{ex}(H)}$ for every connected graph H, provided that Lemma 1 holds for H instead of θ_r. We conjecture that this is the case for every planar graph H. This would also imply that the class of graphs containing H as a minor has the vertex/edge Erdős-Pósa property if and only of H is a planar graph. In our opinion this is one of the more general open questions on the combinatorics of Erdős-Pósa dualities.

References

1. Caprara, A.P.A., Rizzi, R.: Packing cycles in undirected graphs. J. Algorithms **48**(1), 239–256 (2003)
2. Bazgan, C., Tuza, Z., Vanderpooten, D.: Efficient algorithms for decomposing graphs under degree constraints. Discrete Appl. Math. **155**(8), 979–988 (2007)
3. Birmelé, E., Bondy, J.A., Reed, B.A.: The Erdős-Pósa property for long circuits. Combinatorica **27**, 135–145 (2007)
4. Bodlaender, H.L., Fomin, F.V., Lokshtanov, D., Penninkx, E., Saurabh, S., Thilikos, D.M.: (Meta) kernelization. In: Proceedings of the 2009 50th Annual, IEEE Symposium on Foundations of Computer Science (FOCS), pp. 629–638. IEEE Computer Society, Washington, DC (2009)
5. Bodlaender, H.L., Fomin, F.V., Lokshtanov, D., Penninkx, E., Saurabh, S., Thilikos, D.M.: (Meta) kernelization. CoRR, abs/0904.0727 (2009)
6. Bodlaender, H.L., Thomassé, S., Yeo, A.: Kernel bounds for disjoint cycles and disjoint paths. Theoret. Comput. Sci. **412**(35), 4570–4578 (2011)
7. Chatzidimitriou, D., Raymond, J-F., Sau, I., Thilikos, D.M.: Minors in graphs of large θ_r-girth. CoRR, abs/1510.03041 (2015)
8. Chatzidimitriou, D., Raymond, J-F., Sau, I., Thilikos, D.M.: An $o(\log \text{OPT})$-approximation for covering and packing minor models of the pumpkin. CoRR, abs/1510.03945 (2015)
9. Chekuri, C., Chuzhoy, J.: Large-treewidth graph decompositions and applications. In: Proceedings of the Forty-Fifth Annual ACM Symposium on Theory of Computing (STOC), pp. 291–300. ACM, New York (2013)
10. Diestel, R.: Graph Theory. Graduate Texts in Mathematics, 3rd edn. Springer, Heidelberg (2005)
11. Diestel, R., Kawarabayashi, K.I., Wollan, P.: The Erdős-Pósa property for clique minors in highly connected graphs. J. Comb. Theor. Series B **102**(2), 454–469 (2012)
12. Erdős, P., Pósa, L.: On independent circuits contained in a graph. Can. J. Math. **17**, 347–352 (1965)
13. Fiorini, S., Joret, G., Pietropaoli, U.: Hitting diamonds and growing cacti. In: Eisenbrand, F., Shepherd, F.B. (eds.) IPCO 2010. LNCS, vol. 6080, pp. 191–204. Springer, Heidelberg (2010)
14. Fiorini, S., Joret, G., Sau, I.: Optimal Erdős-Pósa property for pumpkins. Manuscript (2013)
15. Fiorini, S., Joret, G., Wood, D.R.: Excluded forest minors and the erdős-pósa property. Comb. Probab. Comput. **22**(5), 700–721 (2013)
16. Fomin, F.V., Lokshtanov, D., Misra, N., Saurabh, S.: Planar F-deletion: approximation, kernelization and optimal FPT-algorithms. In: IEEE 53rd Annual Symposium on Foundations of Computer Science (FOCS), pp. 470–479, October 2012

17. Jim, G., Kasper, K.: The Erdős-Pósa property for matroid circuits. J. Comb. Theor. Ser. B **99**(2), 407–419 (2009)

18. Golovach, P.A.: Personal communication (2015)

19. Joret, G., Paul, C., Sau, I., Saurabh, S., Thomassé, S.: Hitting and harvesting pumpkins. SIAM J. Discr. Math. **28**(3), 1363–1390 (2014)

20. Kakimura, N., Kawarabayashi, K.I., Kobayashi, Y.: Erdős-pósa property and its algorithmic applications: Parity constraints, subset feedback set, and subset packing. In: Proceedings of the Twenty-third Annual ACM-SIAM Symposium on Discrete Algorithms (SODA), pp. 1726–1736. SIAM (2012)

21. Kawarabayashi, K.I., Kobayashi, Y.: Edge-disjoint odd cycles in 4-edge-connected graphs. In: 29th International Symposium on Theoretical Aspects of Computer Science, (STACS), February 29th – March 3rd, 2012, Paris, France, pp. 206–217 (2012)

22. Kawarabayashi, K.-I., Nakamoto, A.: The erdős-pósa property for vertex- and edge-disjoint odd cycles in graphs on orientable surfaces. Discr. Math. **307**(6), 764–768 (2007)

23. Kloks, T., Lee, C.M., Liu, J.: New algorithms for k-face cover, k-feedback vertex set, and k-disjoint cycles on plane and planar graphs. In: Kučera, L. (ed.) WG 2002. LNCS, vol. 2573, pp. 282–295. Springer, Heidelberg (2002)

24. Král', D., Voss, H.-J.: Edge-disjoint odd cycles in planar graphs. J. Comb. Theor. Ser. B **90**(1), 107–120 (2004)

25. Krivelevich, M., Nutov, Z., Salavatipour, M.R., Yuster, J.V., Yuster, R.: Approximation algorithms and hardness results for cycle packing problems. ACM Trans. Algorithms **3**(4) (2007)

26. Krivelevich, M., Nutov, Z., Yuster, R.: Approximation algorithms for cycle packing problems. In: Proceedings of the Sixteenth Annual ACM-SIAM Symposium on Discrete Algorithms (SODA), pp. 556–561. SIAM, Philadelphia, PA (2005)

27. Pontecorvi, M., Wollan, P.: Disjoint cycles intersecting a set of vertices. J. Comb. Theor. Ser. B **102**(5), 1134–1141 (2012)

28. Rautenbach, D., Reed, B.: The Erdős-Pósa property for odd cycles in highly connected graphs. Combinatorica **21**, 267–278 (2001)

29. Raymond, J.-F., Sau, I., Thilikos, D.M.: An edge variant of the Erdős-pósa property. CoRR, abs/1311.1108 (2013)

30. Reed, B.A., Robertson, N., Seymour, P.D., Thomas, R.: Packing directed circuits. Combinatorica **16**(4), 535–554 (1996)

31. Robertson, N., Seymour, P.D.: Graph minors. v. excluding a planar graph. J. Comb. Theor. Ser. B **41**(2), 92–114 (1986)

32. Salavatipour, M.R., Verstraete, J.: Disjoint cycles: integrality gap, hardness, and approximation. In: Jünger, M., Kaibel, V. (eds.) IPCO 2005. LNCS, vol. 3509, pp. 51–65. Springer, Heidelberg (2005)

Submodular Function Maximization on the Bounded Integer Lattice

Corinna Gottschalk$^{(\boxtimes)}$ and Britta Peis

School of Business and Economics, RWTH Aachen University, Aachen, Germany
{corinna.gottschalk,britta.peis}@oms.rwth-aachen.de

Abstract. We consider the problem of maximizing a submodular function on the bounded integer lattice. As a direct generalization of submodular set functions, $f : \{0, \ldots, C\}^n \to \mathbb{R}_+$ is submodular, if $f(x) + f(y) \geq f(x \wedge y) + f(x \vee y)$ for all $x, y \in \{0, \ldots, C\}^n$ where $\wedge$ and $\vee$ denote element-wise minimum and maximum. The objective is finding a vector x maximizing $f(x)$. In this paper, we present a deterministic $\frac{1}{3}$-approximation using a framework inspired by [2]. We also provide an example that shows the analysis is tight and yields additional insight into the possibilities of modifying the algorithm. Moreover, we examine some structural differences to maximization of submodular set functions which make our problem harder to solve.

1 Introduction

Recall that a set function $f : 2^{\mathcal{N}} \to \mathbb{R}$ is called submodular if $f(U) + f(W) \geq f(U \cap W) + f(U \cup W)$ for all $U, W \subseteq \mathcal{N}$. Optimization problems with submodular objective functions have received a lot of attention in recent years. Submodular objectives are motivated by the principle of economy of scale, and thus find many applications in real-world problems. Moreover, submodular functions play a major role in combinatorial optimization. Several combinatorial optimization problems have some underlying submodular structure, for example, cuts in graphs and hypergraphs, or rank functions of matroids.

As a breakthrough result, the problem to find a subset $S \subseteq \mathcal{N}$ minimizing a submodular function f has been shown to be solvable in strongly polynomial time in [11]. In contrast, the corresponding maximization problem

$$\max\{f(S) : S \subseteq \mathcal{N}\} \tag{1}$$

for a nonnegative submodular function f is easily seen to be *NP*-hard, as it contains, for example, MAX CUT as a special case. We refer to (1) as UNCONSTRAINED SUBMODULAR MAXIMIZATION (USM). For USM, Buchbinder et al. presented a deterministic $\frac{1}{3}$-approximation and a randomized $\frac{1}{2}$-approximation in [2]. Both algorithms use a "Double Greedy" framework that starts with two different sets and, for a fixed order of the elements, decides in each step which of the two sets should be modified using the given element. On the other hand, Feige et al. [5] showed that no approximation ratio better than $\frac{1}{2}$ can be achieved.

© Springer International Publishing Switzerland 2015
L. Sanità and M. Skutella (Eds.): WAOA 2015, LNCS 9499, pp. 133–144, 2015.
DOI: 10.1007/978-3-319-28684-6_12

We consider a generalization of submodular set functions: Submodular functions on a subset of the integer lattice $\mathbb{Z}^n$. For $x, y \in \mathbb{Z}^n$ let $(x \vee y)_e$ denote $\max\{x_e, y_e\}$ and $(x \wedge y)_e$ denote $\min\{x_e, y_e\}$ for $e \in \{1, \ldots, n\}$. Then, a function $f : D \to \mathbb{R}$ on a finite set D of the form $D = \{x \in \mathbb{Z}^n | l_i \leq x_i \leq u_i \; \forall i \in \{1, \ldots, n\}\}$ is called submodular if

$$f(x) + f(y) \geq f(x \vee y) + f(x \wedge y) \; \forall x, y \in D.$$

Clearly, this captures submodular set functions since vectors with entries 0 and 1 can be seen as incidence vectors of sets and in that case $\wedge$ and $\vee$ correspond to intersection and union. D is called *bounded integer lattice*.

Submodular functions on the integer lattice have been studied before, for example, in discrete convex analysis, $L^\natural$-convex functions are of this type. We provide more details in "Related Work".

Submodular Maximization on Integer Lattices. Given a bounded integer lattice $D = \{x \in \mathbb{Z}^n | l_i \leq x_i \leq u_i \; \forall i \in \{1, \ldots, n\}\}$ and a submodular function $f : D \to \mathbb{R}_+$, we consider the problem of maximizing f on D:

$$\max\{f(x) : x \in D\}. \tag{2}$$

We will refer to (2) as SUBMODULAR MAXIMIZATION ON A BOUNDED INTEGER LATTICE (SMBIL). For ease of notation, we will from now on assume that $l_i = 0$ and $u_i = C \; \forall i \in \{1, \ldots, n\}$. Thus, we prove all results for a bounded integer lattice of the form $\{0, 1, \ldots, C\}^n$, but all results in this paper can be easily generalized to any bounded integer lattice as defined above.

As mentioned before, SMBIL generalizes USM, thus the hardness of approximation holds as well.

There is another way to interpret the bounded integer lattice as well: Let $x \in \mathbb{Z}^n_+$ denote the incidence vector of a multiset where entries in x specify the multiplicity of individual elements. Then SMBIL can be seen as maximizing a submodular function on multisets containing at most C copies of each element which gives rise to possible applications. Consider for example, a version of the sensor deployment problem presented in [9] where we also can decide how many sensors we want to deploy at a location, perhaps operating under the assumption that a sensor can malfunction and thus backup sensors at vital points make sense.

Our Contribution. Interestingly, SMBIL turns out to be considerably harder than USM. In contrast, minimization of submodular functions on the bounded integer lattice can be solved in pseudopolynomial time since the bounded integer lattice is a distributive lattice over which a submodular function can be minimized efficiently (see e.g. [7]).

As our main result, we present a pseudopolynomial $\frac{1}{3}$-approximation for SMBIL which is tight in the sense that there exists an instance for which the performance ratio is achieved. Notice that the pseudopolynomial running time depends on C, i.e. in a multiset interpretation of SMBIL, it depends on the maximal possible number of elements in a multiset.

Our algorithm generalizes the Double Greedy approach from [2], but the tightness example we provide for SMBIL is in some sense stronger than the one presented for the algorithm in [2]: While that example relies on a specific bad order in which the elements are processed, our example remains tight even if the order of elements is not prescribed but instead the next element to be processed is the one that improves the objective function the most.

While a randomized version of the Double Greedy provides a $\frac{1}{2}$-approximation for the Boolean case, the question whether we can achieve that approximation guarantee for SMBIL remains open. We discuss several approaches for randomizing our algorithm and identify the underlying structures that render SMBIL considerably harder than USM. For example, [5] relies upon a property of local maxima for submodular functions which does not hold for SMBIL.

Related Work. The study of USM goes back to the sixties and has occupied researchers ever since. A comprehensive study on USM has been done by Feige, Mirrokni and Vondrák in [5] who provide and analyze several constant approximation algorithms for USM. In particular, they present a simple Local Search algorithm that yields a $\frac{1}{3}$-approximation. Using a noisy version of f as objective function, they could improve the performance guarantee to $\frac{2}{5}$. Feige et al. in [5] also showed that we cannot hope for a performance guarantee lower than $\frac{1}{2}$. They could prove that any $\frac{1}{2} + \varepsilon$ -approximation would require an exponential number of queries to the oracle. For symmetric submodular functions, they already show that the ratio is tight. Subsequently, Oveis Gharan and Vondrak [14] and Feldmann, Naor and Schwartz [6] improved the approximation ratio to 0.41 and 0.42 respectively and finally Buchbinder, Feldmann, Naor and Schwartz closed the gap and gave a randomzied $\frac{1}{2}$-approximation for USM in [2]. Recently, Buchbinder and Feldman ([3]) also showed how to derandomize this algorithm to obtain a deterministic $\frac{1}{2}$-approximation.

Inaba et al. ([12]) examined the problem of maximizing monotone submodular functions subject to a knapsack constraint on the bounded integer lattice and obtained an approximation guarantee of $1 - \frac{1}{e}$, which is identical to the guarantee for set functions. Soma and Yoshida considered maximizing monotone submodular functions that satisfy an additional property subject to various constraints in ([16]).

While our main interest in SMBIL is of a theoretical nature with the ultimate goal to gain a better understanding of submodular function maximization, there are applications where submodular functions on the integer lattice play a crucial role. For example, Bolandnazar et al. ([1]) showed that Assemble-to-Order Systems for supply chain management can be optimized by minimizing a submodular function on the integer lattice. [12] discusses several applications of their problem as well, e.g. budget allocation and sensor placement.

In discrete convex analysis, several special cases of submodular functions on the integer lattice have been investigated. For example, $L^\natural$-convex and $M^\natural$-concave functions are submodular and $M^\natural$-concave functions can be maximized in polynomial time ([13]).

Preliminaries. As usual, we assume that a submodular function f is given by an oracle. Moreover, Feige et al. mention in [5] that for general submodular functions, it is NP-hard to decide whether there exists a set S such that $f(S) > 0$. Since our main goal is finding good approximation algorithms, restricting to non-negative submodular functions makes sense.

For a vector $x \in \mathbb{Z}^n$ we denote by $(x|x_j = k)$ the vector where all but the entry x_j remain the same and x_j is set to k. As usual, $[n]$ denotes the set $\{0, 1, \ldots, n\}$ and $e^i \in \{0,1\}^n$ denotes the vector where $e_i^i = 1$ and $e_j^i = 0$ for all $j \neq i$.

2   Approximating SMBIL

In this section, we first present a $\frac{1}{3}$-approximation for SMBIL which is inspired by [2]. Then, we show that our analysis is tight and discuss approaches for a randomized Double Greedy algorithm. "Double Greedy" refers to the idea that we start with two vectors a and b and modify them until both vectors are equal, while ensuring f never decreases. In the beginning $a_i = 0$ and $b_i = C$ $\forall i \in \{1, \ldots, n\}$, i.e. initially, a (resp. b) is the unique minimal (maximal) element in D. Now, we traverse the components in a fixed order: For a given index i, we change a_i and b_i while maintaining $a \leq b$ without decreasing the submodular function f.

In the scenario where the integer lattice is bounded by 1, our Algorithm 1 coincides to the one in [2] when vectors are interpreted as characteristic vectors of sets.

Theorem 1. *Let $f : [C]^n \to \mathbb{R}_+$ be a nonnegative submodular function. Then Algorithm 1 is a $\frac{1}{3}$-approximation for SMBIL and has running time $\mathcal{O}(CnT)$ where T is the time for one call to the oracle representing f.*

2.1   Proof of Theorem 1

The proof of Theorem 1 relies upon two lemmas. First, we show that Algorithm 1 really is a Greedy algorithm in the sense that f never decreases:

Lemma 1. *Let $f : [C]^n \to \mathbb{R}_+$ be a submodular function. Then for all $1 \leq i \leq 2n$ and vectors a^i, b^i as in Algorithm 1 the following holds: $f(a^i) \geq f(a^{i-1})$ and $f(b^i) \geq f(b^{i-1})$.*

Proof. Let k be the component in which the vectors a^{i-1} and b^{i-1} will be changed in an iteration of the loop.

By definition, $\delta_{a,i}$ and $\delta_{b,i}$ are nonnegative. Suppose that a is changed first, i.e. $\delta_{a,i} \geq \delta_{b,i}$, then clearly, the lemma holds for a^{i-1} and a^i and $b^i = b^{i-1}$. Now we show that the change from b^i to b^{i+1} does not lead to a decrease in f: We have $f(b^{i-1}|b_k^{i-1} = a_k^i) + f(a^{i-1}|a_k^{i-1} = C) \geq f(a^{i-1}|a_k^{i-1} = a_k^i) + f(b^{i-1})$ by submodularity of f.

Algorithm 1. Generalized Double Greedy for SMBIL

Input: A bounded integer lattice defined by a bound C and a dimension n, a nonnegative submodular function f on $[C]^n$ **Output:** A vector $a \in [C]^n$ Choose an order $l_1, \ldots, l_n$ of the n components of the vectors. Set $a_j^0 = 0, b_j^0 = C \; \forall 1 \le j \le n, i = 1$

for $k = 1$ **to** n **do**

$\quad \delta_{a,i} = \max_{c \in [C]} f(a^{i-1} | a_{l_k}^{i-1} = c) - f(a^{i-1})$

$\quad \delta_{b,i} = \max_{c \in [C]} f(b^{i-1} | b_{l_k}^{i-1} = c) - f(b^{i-1})$

$\quad$ **if** $\delta_{a,i} \ge \delta_{b,i}$ **then**

$\quad\quad$ Let c' be the maximal number among those for which $\delta_{a,i}$ is obtained.

$\quad\quad a^i = (a^{i-1} | a_{l_k}^{i-1} = c'), b^i = b^{i-1}$

$\quad\quad a^{i+1} = a^i, b^{i+1} = (b^i | b_{l_k}^i = c')$

$\quad$ **else**

$\quad\quad$ Let c' be the minimal number among those for which $\delta_{b,i}$ is obtained.

$\quad\quad a^i = a^{i-1}, b^i = (b^{i-1} | b_{l_k}^{i-1} = c')$

$\quad\quad b^{i+1} = b^i, a^{i+1} = (a^i | a_{l_k}^i = c')$

$\quad i = i + 2$

return a^{2n}

As $f(a^{i-1} | a_k^{i-1} = a_k^i) \ge f(a^{i-1} | a_k^{i-1} = c) \; \forall \; a_k^{i-1} \le c \le C$, the above implies

$$f(b^{i-1} | b_k^{i-1} = a_k^i) - f(b^{i-1}) \ge f(a^{i-1} | a_k^{i-1} = a_k^i) - f(a^{i-1} | a_k^{i-1} = C) \ge 0.$$

Since $b^{i-1} = b^i$ and $b^{i+1} = (b^{i-1} | b_k^{i-1} = a_k^i)$, it follows that $f(b^{i+1}) \ge f(b^i)$ as desired.

Moreover, defining b^{i+1} as in the algorithm also yields the biggest increase in f: Assume there exists $c > a_k^i$ such that $f(b^i | b_k^i = c) > f(b^i | b_k^i = a_j^i)$. Then, by submodularity,

$$0 > f(b^i | b_k^i = a_k^i) - f(b^{i-1} | b_k^i = c) \ge f(a^i) - f(a^i | a_k^i = c),$$

which contradicts the definition of $\delta_{a,i}$ as the maximal increase in f for a^{i-1}. The case where $\delta_{b,i} > \delta_{a,i}$ can be shown in the same way.   □

Let OPT denote a fixed optimal solution for SMBIL. In order to bound the value of a solution of Algorithm 1 with respect to the optimum, we define $OPT^i = (OPT \vee a^i) \wedge b^i$ analogous to [2]. Consequently, $OPT^0 = OPT$ and $OPT^{2n} = a^{2n} = b^{2n}$. In Lemma 2, we show $f(OPT^{i-1}) - f(OPT^i) \le f(a^i) - f(a^{i-1}) + f(b^i) - f(b^{i-1})$, i.e. the increase in OPT^i is bounded for each change of the vectors a or b.

Then, we can use Lemma 2 to prove Theorem 1: $f(OPT^0) - f(OPT^{2n}) = \sum_{i=1}^{2n} (f(OPT^{i-1}) - f(OPT^i)) \le \sum_{i=1}^{2n} \left(f(a^i) - f(a^{i-1}) \right) + \sum_{i=1}^{2n} \left(f(b^i) - f(b^{i-1}) \right) = f(a^{2n}) + f(b^{2n}) - f(a^0) - f(b^0) \le f(a^{2n}) + f(b^{2n})$. This is equivalent to $f(OPT) \le 3f(a^{2n})$ and the algorithm returns a^{2n}. Since the running time is obvious, this concludes the proof of Theorem 1 except for Lemma 2:

Lemma 2. *For a submodular function $f : [C]^n \to \mathbb{R}_+$ the following holds in Algorithm 1 for all $1 \leq i \leq 2n$ where $OPT^i := (OPT \vee a^i) \wedge b^i$:*

$$f(OPT^{i-1}) - f(OPT^i) \leq f(a^i) - f(a^{i-1}) + f(b^i) - f(b^{i-1}).$$

Proof. Let us consider the case where $a^i \neq a^{i-1}$ and let k be the index, where the vectors differ. If $a_k^i \leq OPT_k$ then $OPT_k^{i-1} = OPT_k^i$ since $b_k^i = b_k^{i-1}$ and thus $OPT^{i-1} = OPT^i$. Since Lemma 1 implies that the right-hand side of the equation is nonnegative, the inequality holds.

So we assume now that $a_k^i > OPT_k$. Since $b_k^i \geq a_k^i$, we have $OPT_k^i = a_k^i$, all other entries of OPT^{i-1} remain unchanged. Moreover, $a_k^{i-1} = 0$, so $OPT_k^{i-1} = OPT_k$.

Submodularity of f implies

$$f(OPT^i) + f(b^{i-1}|b_k^{i-1} = OPT_k) \geq f(OPT^{i-1}) + f(b^{i-1}|b_k^{i-1} = a_k^i) \quad (3)$$
$$\Leftrightarrow f(OPT^i) + f(b^{i-1}|b_k^{i-1} = OPT_k) - f(b^{i-1})$$
$$\geq f(OPT^{i-1}) + f(b^{i-1}|b_k^{i-1} = a_k^i) - f(b^{i-1}).$$

Suppose that $f(OPT^{i-1}) - f(OPT^i) > f(a^i) - f(a^{i-1})$, otherwise we are done.

Using this assumption and (3) yields

$$f(b^{i-1}|b_k^{i-1} = OPT_k) - f(b^{i-1}) + f(OPT^{i-1}) - (f(a^i) - f(a^{i-1}))$$
$$> f(b^{i-1}|b_k^{i-1} = OPT_k) - f(b^{i-1}) + f(OPT^i)$$
$$\geq f(OPT^{i-1}) + f(b^{i-1}|b_k^{i-1} = a_k^i) - f(b^{i-1}).$$

But we have $f(a^i) - f(a^{i-1}) \geq f(b^{i-1}|b_k^{i-1} = c) - f(b^{i-1}) \; \forall \; a_k^{i-1} \leq c \leq b_k^{i-1}$. This is true by design of the algorithm for the first change in an iteration of the loop. For the second change in an iteration, the other vector cannot improve further, so the claim holds as well.

This implies

$$0 \geq f(b^{i-1}|b_k^{i-1} = OPT_k) - f(b^{i-1}) - (f(a^i) - f(a^{i-1})) \quad (4)$$
$$> f(b^{i-1}|b_k^{i-1} = a_k^i) - f(b^{i-1}).$$

But in the proof of Lemma 1, we have shown that $f(b^{i-1}|b_k^{i-1} = a_k^i) - f(b^{i-1}) \geq 0$, so (4) is a contradiction.

The case $b^i \neq b^{i-1}$ can be treated analogously and if neither case applies, then clearly $OPT^i = OPT^{i-1}$ as well. $\square$

2.2   The Guarantee of $\frac{1}{3}$ is tight

Since Algorithm 1 generalizes the deterministic algorithm in [2], the tight example they provide also works for our algorithm. But our example has a few additional properties: First, even if we do not prescribe the order of the components in advance and instead choose the index that results in the biggest increase in f for each step, the algorithm does not yield a better solution (see Theorem 2 below) which is not true for the example provided in [2]. Second, we will show in the next section that our example also implies that an analogue to the analysis of the randomized algorithm in [2] is not possible for bounded integer lattices.

Theorem 2. *For an arbitrarily small constant $\varepsilon > 0$ there exists a submodular function for which Algorithm 1 provides only a $\frac{1}{3} + \varepsilon$-approximation. This result still holds if the components are ordered such that at any time the chosen component yields the biggest increase in f.*

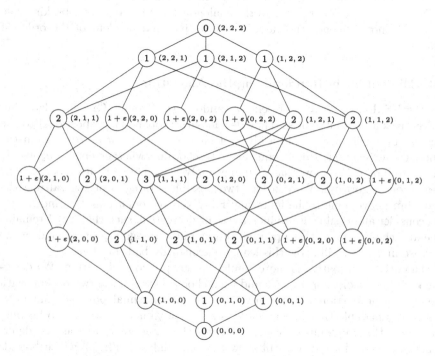

Fig. 1. Each node corresponds to a vector in $\{0, 1, 2\}^3$, written next to it, with the value of f within the node. The picture can be read similar to Hasse diagrams of posets: There is a line connecting two vectors, if their L_1-distance equals 1. If we think of the edges as being directed upwards, then the meet of two vectors is their biggest common predecessor, and their join the smallest common successor.

Proof. Consider the following submodular function $f : \{0, 1, 2\}^3 \to \mathbb{R}_+$. Let $f'(a) = \min\{|\{a_i : a_i > 0\}|, |\{a_i : a_i < 2\}|\}$. Now we define $f(a) = 1 + \varepsilon$ if a consists of either the entries $2, 0, 2$ or $2, 0, 0$ in any order. We set $f(2, 1, 0) = f(0, 1, 2) = 1 + \varepsilon$. For all other vectors, we set $f(a) = f'(a)$. It can be checked that f is indeed submodular. To give a better intuition, Fig. 1 illustrates this function.

We analyze Algorithm 1 for this instance: Obviously, the optimum is the vector $(1, 1, 1)$ of value 3, but depending on the order in which the entries of the vectors are processed, Algorithm 1 may terminate with a set of value $1 + \varepsilon$: Let l_1, l_2, l_3 be the predetermined order. In the first iteration, the maximal possible gain in f is $1 + \varepsilon = \delta_{a,1} = \delta_{b,1}$. So we we set $a_{l_1}^1 = 2$ and $b_{l_1}^2 = 2$. Next, $\delta_{b,3} = 1 + \varepsilon > \delta_{a,3}$ and thus $b_{l_2}^3 = 0$ and $a_{l_2}^4 = 0$. What happens now depends on the chosen order of indices, consider $1, 3, 2$. Then $a^4 = (2, 0, 0)$ and $b^4 = (2, 2, 0)$, both of value $1 + \varepsilon$ and the algorithm returns the vector $a^6 = b^6 = (2, 2, 0)$ of value $1 + \varepsilon$. Since the maximal possible gain in f is $1 + \varepsilon$ for the first two indices we consider, independent of the order, the same order could have been chosen by an algorithm that always processes the index which maximizes the gain in f.

Notice that the order $3, 1, 2$ works analogously and that the tie-braking rule for δ does not influence the value of the output, independent of the order of indices.   □

2.3   Difficulties in Randomizing the Algorithm

For the USM case, [2] also presents a randomized "Double Greedy" algorithm which gives a guarantee of $\frac{1}{2}$: They decide with probability proportional to the increase in f whether to add a given element to one set or delete it from the other. In our context, this is equivalent to choosing whether entries a_i and b_i are set to 0 or 1. We show that a similar analysis cannot work if we adapt their idea to our algorithm. We consider two possible strategies of generalizing the algorithm above. One is the following: For given vectors a, b and an index i we can consider all possibilities to increase a_i or decrease b_i such that $a \leq b$ remains true and choose one of these possibilities at random (again proportional to the increase in f). We will show this leads to arbitrarily bad solutions.

The other alternative is more similar to our previous algorithm: We determine the best choice for a and b and then choose between the two options with probability proportional to the δ-values, i.e. the maximal possible gain in f. While it is possible that this actually is a $\frac{1}{2}$-approximation, we can so far only show that this randomized algorithm gives the same guarantee as the deterministic version. Indeed, we will show that an analysis as in [2] which bounds the expected decrease of $f(OPT^i)$ in each step cannot prove a guarantee better than $\frac{1}{3}$ by examining the tight example for the deterministic algorithm which was given in Sect. 2.2. So, while it might still be true that this randomized version of our algorithm actually is a $\frac{1}{2}$-approximation (for example, the worst case expected value in the given example is $1.5 + \varepsilon$), we would require an analysis that takes a more global view.

We should note that for the Boolean lattice, both these approaches are identical and correspond to the randomized algorithm in [2].

Randomized Choice over All Possible Solutions

Lemma 3. *Consider the following randomized version of Algorithm 1: Assume an order in which we iterate over the indices is given, consider index i. While*

$a_i < b_i$, *randomly set either* $a = (a|a_i = c)$ *or* $b = (b|b_i = c)$ *where the choice is made over all* $a_i \leq c \leq b_i$ *proportional to the increase in* f *if it is positive. If the increase in* f *is non-positive for all possibilities, we arbitrarily choose an option where the increase in* f *is* 0, *but* a *or* b *are changed. This algorithm for SMBIL can perform arbitrarily bad.*

Proof. First, we remark that in the case where the increase in f is non-positive everywhere, a choice as above exists. We now analyze the following instance: Given $C \in \mathbb{N}$, $\varepsilon > 0$, define a submodular function $f : [C]^2 \to \mathbb{R}$ as follows:

$$f(x) = \begin{cases} 0, & \text{if } x = (0,0) \text{ or } x = (C,C) \\ 1 & \text{if } x = (C,0) \\ \varepsilon & \text{else} \end{cases}.$$

Our randomized algorithm starts with $a = (0,0)$ and $b = (C,C)$. We fix an index, say 1. Until $a_1 = b_1$, we change entry a_1 or b_1 and choose a value such that $a_1 \leq b_1$ at random proportional to the increase in f. I.e., for the first step, we have $2C$ options where the increase in f is positive, all but one (setting $a_1 = C$) will be taken with probability $\frac{\varepsilon}{(2C-1)\varepsilon+1}$.

After the first step, either $b_1 < C$ or $a_1 > 0$. In the second case, if $a_1 < C$, the only options now to increase f are changing b_1 or setting $a_1 = C$. Therefore, after two steps either $a_1 = C$ or $b_1 < C$ and depending on which of these holds, the return value will be 1 or ε. Thus, the probability for the algorithm to return a vector of value ε is

$$Pr[b \text{ is changed first}] + \sum_{i=1}^{C-1} Pr[a_1 \text{ is set to } i \text{ first, then } b \text{ is changed}]$$

$$= \frac{C\varepsilon}{(2C-1)\varepsilon+1} + \frac{\varepsilon}{(2C-1)\varepsilon+1} \sum_{i=1}^{C-1} \frac{(C-i)\varepsilon}{(C-i)\varepsilon+1-\varepsilon}$$

Since for $C \to \infty$ this expression converges to 1, the expected return value of the algorithm converges to ε, for details we refer to the extended version. □

Note that variations like randomizing over combinations of values for a_i and b_i and choosing proportional to the sum of the increases in f show a similar behavior.

Randomized Choice over the Best Possible Solutions

Lemma 4. *Consider the following randomized version of Algorithm 1: Instead of deciding to change a^{i-1} or b^{i-1} depending on whether $\delta_{a,i} \geq \delta_{b,i}$, randomly change a or b proportional to $\max\{0, \delta_{a,i}\}$ and $\max\{0, \delta_{b,i}\}$. Then adapt the other vector as before. For this algorithm, there is no constant $c < 1$ such that $E[f(OPT^{i-1}) - f(OPT^i)] \leq c \cdot E[f(a^i) - f(a^{i-1}) + f(b^i) - f(b_{i-1})]$.*

In particular, unlike in [2], an analysis that bounds the expected decrease of OPT^i by the expected increase in a and b in each step can not yield an approximation factor better than $\frac{1}{3}$.

Proof. As before, OPT^i is defined as $(OPT \vee a^i) \wedge b^i$ for a fixed optimal solution OPT. Consider the example presented in the previous section in Fig. 1. No matter how we choose the first index j, $a_j = 2$ or $b_j = 0$ both with equal probability. Thus, OPT^1 consists of the entries $0, 1, 1$ or $2, 1, 1$ and thus has value 2 in both scenarios which implies $E[f(OPT^0) - f(OPT^1)] = \frac{1}{2} \cdot ((3 - 2) + (3 - 2)) = 1$ and $E[f(a^1) - f(a^0) + f(b^1) - f(b_0)] = 2 \cdot \frac{1}{2} \cdot (1 + \varepsilon)$. Note that the statement is true no matter how the order of indices is chosen.   □

3   Structural Differences in SMBIL and USM

There are a number of interesting differences between properties of submodular set functions and of submodular functions on the integer lattice. First of all, note that for $n = 1$ we cannot expect f to have any structure since all functions are submodular for $n = 1$. Therefore, we investigate the case where $n > 1$. We will present a examples to give an intuition what are the challenges of SMBIL compared to USM. First of all, note that for $n = 1$ we cannot expect f to have any structure, since in that case, all functions are submodular. Therefore, we investigate the case where $n > 1$.

The local search approach presented in [5] uses a generalization of the following result (Theorem 3) about local optima. For a submodular function $f : [C]^n \to \mathbb{R}_+$, we follow the definition used in [13] and say that x is a local optimum if $f(x) \geq \max\{f((x - p) \vee \mathbf{0}), f((x + p) \wedge C)\}$ for all $p \in \{0, 1\}^n$. Here, $\mathbf{0}$ (resp. C) denote the vectors consisting of all 0s, (resp. Cs). These ensure that all values are within the domain of f. This definition is more general than the one commonly used on the Boolean lattice, where a local optimum is compared to all sets that can be constructed by adding or deleting one element.

Theorem 3 [8]. *Given a submodular function* $f : \{0, 1\}^n \to \mathbb{R}_+$, *if* x *is a local optimum (w.r.t. the Boolean lattice) of* f *and if* $y \geq x$ *or* $y \leq x$, *then* $f(x) \geq f(y)$.

But the above theorem is no longer true for the integer lattice even for the stronger definition of local optima:

Proposition 1. *There exists a submodular function* $f : [C]^n \to \mathbb{R}_+$, $n > 1$ *with a local optimum* x *and a vector* y *with* $y \geq x$ *such that* $f(y) > f(x)$.

Consider a submodular function $f : \{0, 1, 2\}^2 \to \mathbb{R}^+$ defined as follows: $f(0, 0) = 1, f(1, 0) = 2, f(2, 0) = 1$. Now $f(i, 1) = f(i, 0) - 1$, $f(i, 2) = f(i, 0) + 1$ except for $f(0, 2) = \lambda$ for some $\lambda \geq 2$. This function is submodular and $(1, 0)$ is a local optimum, but $3 = f(1, 2) > 2 = f(1, 0)$. For an illustration, see Fig. 2.

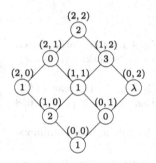

Fig. 2. Each node corresponds to a vector in $\{0,1,2\}^2$, the values in the nodes define a submodular function f for any $\lambda \geq 2$.

Moreover, a simple Greedy algorithm also shows different behaviors for USM and SMBIL:

Proposition 2. *Let* $f : [C]^n \to \mathbb{R}_+$ *be a submodular function. Consider the following Greedy algorithm: Start with the all-zero vector* $a = (0, \dots, 0)$*. While it is possible to not decrease a by increasing an entry by one (i.e. there exists i with $f(a + e^i) \geq f(a)$), choose the best variant and update a (i.e. set $a = a + e^i$ for $i = argmax_{1 \leq i \leq n}(f(a + e^i) - f(a))$). This is an $\mathcal{O}(n)$-approximation for $C = 1$ and arbitrarily bad for $C > 1$.*

In the Boolean case, this is the simplest possible Greedy: We start with the empty set and add the best element as long as that is possible without decreasing f. It is folklore that this is an $\mathcal{O}(n)$-approximation and there are instances where this bound is tight even for the special case of DIRECTED MAX CUT. For SMBIL on the other hand, such an algorithm can be arbitrarily bad, as we can see by examining the previous submodular function. Such a Greedy algorithm will never increase the second entry in the vector from 0, so it can never reach the element of value λ, no matter how large λ is.

4   Discussion and Open Problems

One of the main remaining questions is, of course, whether there is a $1/2$- approximation for SMBIL or whether the problem is harder to approximate than USM. One could rightfully ask why we use a Greedy strategy instead of e.g. using multilinear relaxation introduced in [4] which was successfully applied to USM. Generalizing this to our setting, however, induces several problems. In particular, for the natural generalization of the multilinear relaxation to the domain $[0, C]^n$ the partial derivatives do not exist everywhere.

Furthermore, maximizing a submodular function on distributive lattices is an interesting generalization of SMBIL with connections to algorithmic game theory: In [15] Schulz et al. used USM as a subroutine to approximate the least core in cooperative games with supermodular costs. If we consider this problem for distributive games as introduced by Grabisch et al. ([10]) with supermodular cost functions, a good approximation for maximizing submodular functions on distributive lattices would be helpful. Unfortunately, our results do not easily generalize to submodular functions on distributive lattices.

Acknowledgement. We thank S.Thomas McCormick and Kazuo Murota for fruitful discussions.

References

1. Bolandnazar, M., Huh, W., McCormick, S., Murota, K.: A note on "order-based cost optimization in assemble-to-order systems". University of Tokyo (February, Techical report (2015)
2. Buchbinder, N., Feldmann, M., Naor, J., Schwartz, R.: A tight linear (1/2)-approximation for unconstrained submodular maximization. In: FOCS pp. 649–658 (2012)
3. Buchbinder, N., Feldman, M.: Deterministic algorithms for submodular maximization problems. CoRR (2015). abs/1505.02695
4. Calinescu, G., Chekuri, C., Pál, M., Vondrák, J.: Maximizing a submodular set function subject to a matroid constraint (extended abstract). In: Fischetti, M., Williamson, D.P. (eds.) IPCO 2007. LNCS, vol. 4513, pp. 182–196. Springer, Heidelberg (2007)
5. Feige, M.: Vondrák: Maximizing Non-Monotone Submodular Functions. SIAM Journal on Computing 40(4), 1133–1153 (2011)
6. Feldman, M., Naor, J.S., Schwartz, R.: Nonmonotone submodular maximization via a structural continuous greedy algorithm. In: Aceto, L., Henzinger, M., Sgall, J. (eds.) ICALP 2011, Part I. LNCS, vol. 6755, pp. 342–353. Springer, Heidelberg (2011)
7. Fujishige, S.: Submodular Functions and Optimization. Annals of Discrete Mathematics, vol. 58, 2nd edn, pp. 305–308. Elsevier, New York (2005)
8. Goldengorin, B., Tijssen, G., Tso, M.: The maximization of submodular functions: old and new proofs for the correctness of the dichotomy algorithm. Technical report 99A17, University of Groningenm Research Institute SOM (1999)
9. Golovin, D., Krause, A.: Submodular function maximization. In: Bordeaux, L., Hamadi, Y., Kohli, P. (eds.) Tractability: Practical Approaches to Hard Problems. Cambridge University Press, Cambridge (2014)
10. Grabisch, M., Xie, L.: The restricted core of games on distributive lattices: how to share benefits in a hierarchy. Math. Methods Oper. Res. 73, 189–208 (2011)
11. Grötschel, M., Lovász, L., Schrijver, A.: Geometric Algorithms and Combinatorial Optimization. Springer, Berlin (1988)
12. Inaba, K., Kakimura, N., K., K., Soma, T.: Optimal budget allocation: Theoretical guarantee and efficient algorithm. In: Proceedings of the 31st International Conference on Machine Learning, pp. 351–359 (2014). JMLR.org
13. Murota, K.: Submodular function minimization and maximization in discrete convex analysis. RIMS Kokyuroku Bessatsu B23, 193–211 (2010)
14. Oveis Gharan, S., Vondrák, J.: Submodular maximization by simulated annealing. In: SODA, pp. 1098–1117 (2011)
15. Schulz, A., Uhan, N.: Approximating the least core value and least core of cooperative games with supermodular costs. Discrete Optim. 10(2), 163–180 (2013)
16. Soma, T., Yoshida, Y.: Maximizing submodular functions with the diminishing return property over the integer lattice. CoRR 2015. abs/1503.01218

Geometric Hitting Set for Segments
of Few Orientations

Sándor P. Fekete[1], Kan Huang[2], Joseph S.B. Mitchell[2], Ojas Parekh[3],
and Cynthia A. Phillips[3](✉)

[1] TU Braunschweig, Braunschweig, Germany
s.fekete@tu-bs.de
[2] Stony Brook University, Stony Brook, NY, USA
{khuang,jsbm}@ams.sunysb.edu
[3] Sandia National Labs, Albuquerque, NM, USA
{odparek,caphill}@sandia.gov

Abstract. We study several natural instances of the geometric hitting
set problem for input consisting of sets of line segments (and rays, lines)
having a small number of distinct slopes. These problems model path
monitoring (e.g., on road networks) using the fewest sensors (the "hitting
points"). We give approximation algorithms for cases including (i) lines
of 3 slopes in the plane, (ii) vertical lines and horizontal segments, (iii)
pairs of horizontal/vertical segments. We give hardness and hardness of
approximation results for these problems. We prove that the hitting set
problem for vertical lines and horizontal rays is polynomially solvable.

1 Introduction

The set cover problem is fundamental in combinatorial optimization. It is NP-
hard and has an $O(\log n)$-approximation algorithm, which is best possible (unless
$P = NP$, [13]). Equivalently, set cover can be cast as a hitting set problem:
given a collection, $\mathcal{C}$, of subsets of set U, find a smallest cardinality set $H \subseteq U$
such that every set in $\mathcal{C}$ contains at least one element of H. Numerous special
instances of set cover/hitting set have been studied. Our focus in this paper is on
geometric instances that arise in covering (hitting) sets of (possibly overlapping)
line segments using the fewest points ("hit points"). A closely related problem is
the "Guarding a Set of Segments" (GSS) problem [3,5,6,25], in which the segments may cross arbitrarily, but do not overlap. Since this problem is strongly
NP-complete [5] in general, our focus is on special cases, primarily those in which
the segments come from a small number of orientations (e.g., horizontal, vertical). We provide several new results on hardness and approximation algorithms.

We also are motivated by the path monitoring problem: given a set of trajectories, each a path of line segments in the plane, place the fewest sensors (points)
to observe (hit) all trajectories. To gain theoretical insight into this challeng-
ing problem, we examine cleaner, but progressively harder, versions of hitting
trajectory/line-like objects with points. If the trajectories are on a Manhattan
road network, the paths are (possibly overlapping) horizontal/vertical segments.

© Springer International Publishing Switzerland 2015
L. Sanità and M. Skutella (Eds.): WAOA 2015, LNCS 9499, pp. 145–157, 2015.
DOI: 10.1007/978-3-319-28684-6_13

Alternatively, one wishes to place the fewest vendors or service stations in a road network to service a set of customer trajectories.

Our Results. We give complexity and approximation results for several geometric hitting set problems on inputs S of line "segments" of special classes, mostly of fixed orientations. The segments are allowed to overlap arbitrarily. We consider various cases of "segments" that may be bounded (line segments), semi-infinite (rays), or unbounded in both directions (lines). Our results are:

(1) Hitting lines of 3 slopes in the plane is NP-hard (greedy is optimal for 2 slopes). For set cover with set size at most 3, standard analysis of the greedy algorithm gives an approximation factor of $H(3) = 1+(1/2)+(1/3) = (11/6)$, and there is a 4/3-approximation based on semi-local optimization [15]. We prove that the greedy algorithm in this special geometric case is a (7/5)-approximation.
(2) Hitting vertical lines and horizontal rays is polytime solvable.
(3) Hitting vertical lines and horizontal (even unit-length) segments is NP-hard. Our proof shows hitting horizontal and vertical unit-length segments is also NP-hard. We prove APX-hardness for hitting horizontal and vertical segments.
(4) Hitting vertical lines and horizontal segments has a (5/3)-approximation algorithm. (This problem has a straightforward 2-approximation).
(5) Hitting pairs of horizontal/vertical segments has a 4-approximation. Hitting pairs having one vertical and one horizontal segment has a (10/3)-approximation. These results are based on LP-rounding. More generally, hitting sets of k segments from r orientations has a $(k \cdot r)$-approximation algorithm.
(6) We give (in the full paper) a linear-time combinatorial 3-approximation algorithm for hitting triangle-free sets of (non-overlapping) segments. A 3-approximation for this version of GSS was recently given [25] using linear programming.

Related Work. There is a wealth of related work on geometric set cover and hitting set problems; we do not attempt here to give an exhaustive survey. The *point line cover* (PLC) problem (see [23,27]) asks for a smallest set of lines to cover a given set of points; it is equivalent, via point-line duality, to the hitting problem for a set of lines. The PLC (and thus the hitting problem for lines) was shown to be NP-hard [30]; in fact, it is APX-hard [7] and Max-SNP Hard [28]. The problem has an $O(\log OPT)$-approximation (e.g., greedy – see [26]); in fact, the greedy algorithm for PLC has worst-case performance ratio $\Omega(\log n)$ [16].

Hassin and Megiddo [22] considered hitting geometric objects with the fewest lines having a small number of distinct slopes. They observed that, even for covering with axis-parallel lines, the greedy algorithm has an approximation ratio that grows logarithmically. They gave approximations for the problem of hitting horizontal/vertical segments with the fewest axis-parallel lines (and, more generally, with lines of a few slopes). Gaur and Bhattacharya [19] consider covering points with axis-parallel lines in d-dimensions; they give a $(d-1)$-approximation based on rounding the corresponding linear program (LP) formulation. Many other stabbing problems (find a small set of lines that stab a given set of objects) have been studied; see, e.g., [14, 17, 20, 21, 26, 29].

A recent paper [25] gives a 3-approximation for hitting sets of "triangle-free" segments. Brimkov et al. [3, 5, 6] have studied the hitting set problem on line segments, including various special cases; they refer to the problem as "Guarding a Set of Segments", or GSS. GSS is a special case of the "art gallery problem:" place a small number of "guards" (e.g., points) so that every point within a geometric domain is "seen" by at least one guard [32, 34]. Brimkov et al. [4] provide experimental results for three GSS heuristics, including two variants of "greedy," showing that in practice the algorithms perform well and are often optimal or very close to optimal. They prove, however, that, in theory, the methods do not provide worst-case constant-factor approximation bounds. For the special case that the segments are "almost tree (1)" (a connected graph is an *almost tree (k)* if each biconnected component has at most k edges not in a spanning tree of the component), a $(2 - \varepsilon)$-approximation is known [3].

An important distinction between GSS and our problems is that allow *overlapping* (or partially overlapping) segments (rays, and lines), while, in GSS, each line segment is maximal in the input set of line segments (the union of two distinct input segments is not a segment). A special case of our problem is *interval stabbing* on a line: Given a set of segments (intervals), arbitrarily overlapping on a line, find a smallest hitting set of points that hit all segments. A simple sweep along the line solves this problem optimally: when a segment ends, place a point and remove all segments covered by that point.

If no point lies within three or more objects, then the hitting set problem is an edge cover problem in the intersection graph of the objects. In particular, if no three segments pass through a common point, the problem can be solved optimally in polynomial time. (This implies that in an arrangement of "random" segments, the GSS problem is almost surely polynomially solvable; see [3]).

Hitting axis-aligned rectangles is related to hitting horizontal and vertical segments. Aronov, Ezra, and Sharir [2] provide an $O(\log \log OPT)$-approximation for hitting set for axis-aligned rectangles (and axis-aligned boxes in 3D), by proving a bound of $O(\varepsilon^{-1} \log \log(\varepsilon^{-1}))$ on the ε-net size of the corresponding range space. The connection between hitting sets and ε-nets [8, 11, 12, 18] implies a c-approximation for hitting set if one can compute an ε-net of size c/ε; recent major advances [1, 33] on lower bounds on ε-nets imply that associated range spaces (rectangles and points, lines and points, points and rectangles) have ε-nets of size superlinear in $1/\varepsilon$. Remarkably, improved $(1 + \varepsilon)$-approximation algorithms (i.e., PTASs) for certain geometric hitting set and set cover problems are possible with simple local search. For example, Mustafa and Ray [31] give a local search PTAS for computing a smallest subset, of a given set of disks, that covers a given set of points. Hochbaum and Maas [24] used grid shifting to obtain a much earlier PTAS for the minimum unit disk cover problem when disks can be placed anywhere in the plane, not restricted to a discrete input set.

2   Hitting Segments

Suppose S is a set of line segments in the plane. If all segments are horizontal, then we can compute an optimal hitting set by independently solving the interval stabbing problem along each of the horizontal lines determined by the input.

If the segments are of two different orientations (slopes), then the problem becomes significantly harder. Without loss of generality, assume the segments are horizontal and vertical. We show the problem is hard even if the axis-parallel segments are all the same length. This result (Corollary 1) is a consequence of an even stronger result, Theorem 4, which we establish in Sect. 5.

We get an immediate 2-approximation algorithm by solving optimally each of the two orientations, and using the union of the hitting points for both. (This generalizes to a k-approximation for hitting sets of segments of k orientations).

3   Hitting Lines

When S is a set of n lines in the plane, greedy gives an $O(\log OPT)$ approximation factor; any approximation factor better than logarithmic would be quite interesting. (See [16,27].) If the lines have only 2 slopes, then greedy is optimal.

3.1   Hardness of Hitting Lines of 3 Slopes in 2D

We prove that the hitting set problem is NP-hard when lines have more than two orientations. Consider the dual formulation: (3-SLOPE-LINE-COVER, 3SLC) Find a minimum-cardinality set of non-vertical lines to cover a set P of points (duals to the set S of lines), which are known to lie on three vertical lines.

We prove (in the full paper) that 3SLC is NP-hard from 3-SAT, using variable gadgets and clause gadgets that rely on carefully placed points on three vertical lines. "Propagation" of variable assignments is determined by triples of points on distinct vertical lines coverable by a single line.

Theorem 1. *The problem 3SLC is NP-complete.*

3.2   Analysis of the Greedy Hitting Set Algorithm for Lines of 3 Slopes in 2D

If no point lies in more than k sets, the greedy algorithm's approximation factor is $H(k) = \sum_{i=1}^{k}(1/i)$ [10]. This property holds for lines of 3 slopes with $k = 3$, giving a greedy approximation factor $H(3) = 11/6$. We give a new analysis, exploiting the special geometric structure of the hitting set problem for lines of 3 slopes, to obtain an approximation factor $(7/5)$; see the full paper.

3.3   Axis-Parallel Lines in 3D

While in 2D the hitting set problem for axis-parallel lines is easily solved, in 3D we prove (in the full paper) that the corresponding hitting set problem is NP-hard, using a reduction from 3-SAT.

Theorem 2. *Hitting set for axis-parallel lines in 3D is NP-complete.*

4   Hitting Rays and Lines

Hitting rays is "harder" than hitting lines, since any instance of hitting lines has a corresponding equivalent instance as a hitting rays problem (place the apices of the rays far enough away that they are effectively lines). Unlike lines, there can be many different collinear rays. Divide collinear rays into two groups according to the direction they point along the containing line, ℓ; because of nesting, we need keep only one of the rays pointing in each of the two directions along ℓ.

We show that the special case with horizontal rays and vertical lines (abbreviated HRVL) is exactly solvable in polynomial time:

Theorem 3. *The hitting set problem for vertical lines and horizontal rays can be solved in $O(nT)$ time, where n is the number of entities and T is the time for computing a maximum matching in a bipartite graph with n nodes.*

We begin with a high-level overview of the algorithm. A point can cover at most 3 objects: a vertical line, a left-facing ray, and a right-facing ray. This requires the two rays to intersect in a segment, and the vertical line to intersect this segment. We call these points 3-hitters. We can compute the maximum possible number of 3-hitters via maximum matching in a bipartite graph, where edges represent intersections between vertical lines and horizontal segments. We prove there exists an optimal solution with this maximum number of 3-hitters. The algorithm performs a sweep inward from the left and right, finding a suitable set of 3-hitters, ensuring the remaining lines have the best possible chance to share a point with the remaining rays. Once everything that is 3-hit is removed, the remaining objects intersect in at most pairs. So we can finish the hitting by solving an edge cover problem between rays and lines. We prove this is optimal.

We now give additional algorithmic and proof details. We call a horizontal ray to the left (resp., right) an *l-ray* (resp. *r-ray*). In this section, all lines are vertical. If two collinear rays are disjoint, we shift one ray slightly up or down, so no two disjoint rays are collinear. These rays cannot be covered by a single point, so this does not fundamentally alter the optimal solution.

If a line only contains one ray, we add a ray to pair with it. For example, if an r-ray intersects no l-ray, we add an intersecting l-ray whose right endpoint is to the right of all vertical input lines. This additional ray won't change the optimal solution. If an l-ray and r-ray intersect, their intersection is a segment. Since all rays intersect another ray, we represent each pair of rays by their segment.

Let H and V denote the number of segments and lines respectively. Any solution requires V points to cover the lines. Those points can help "hit" segments in two possible ways: (1) Place a point on the segment. We call the corresponding line a 3-hitter and say the segment is 3-hit by the line. (2) Hit each ray outside its intersecting segment. This requires two points. We call the left(right) line an l-hitter(r-hitter). We say the segment is *double-hit* by those two lines.

Let v_1 and v_2 be the number of segments hit by the V points in the first and second ways respectively. Then the number of points in the solution is $H + V - v_1 - v_2$. We must put a point on each line to hit as many segments as possible.

Given an instance of HRVL, we can calculate the maximum number of 3-hitters. We construct a bipartite graph G where one set of nodes is the lines and the other set of nodes is the segments. There is an edge between two nodes if and only if the line and segment they represent intersect. Maximum matching in a bipartite graph is solvable in polynomial time. A matching in the graph represents a set of independent intersections in the corresponding HRVL. That is, a set of M edges in a matching corresponds to a way to cover M segments and M lines with M points. These are coverages of type 1. The following intuitive lemma shows it is better to adopt the first way to hit segment.

Lemma 1. *For any instance of HRVL, there is a maximum matching between lines and segments that can be augmented to be an optimal solution.*

Proof Sketch. We use contradiction. Let v_1^* be the largest v_1 for any minimum hitting set. We assume that v_1^* is less than m, the cardinality of the maximum matching between lines and segments. Thus, there is an augmenting path in the bipartite graph G, such as the green path in Fig. 1. Because the current solution is optimal, any augmenting path cannot improve it. This allows us to infer some properties of the first segment and the last line on the augmenting path. We consider the augmenting path P with the shortest length and the shortest horizontal distance between the last two lines. Then by case analysis on path P, we argue there exists another augmenting path that increases v_1^* or violates a minimality condition of P. The proof appears in the full paper.

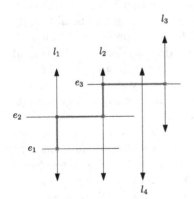

Fig. 1. A green augmenting path: the matching size increases by replacing blue circles with red crosses (Color figure online).

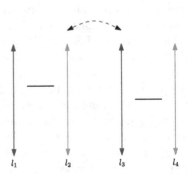

Fig. 2. Swapping l_2 and l_3 makes both of them more useful (Color figure online).

Lemma 2. *Given an optimal solution $\mathcal{S}$, there is an optimal solution $\mathcal{S}'$ that has the same set of 3-hitters as $\mathcal{S}$, with its l-hitters all left of its r-hitters.*

Proof. In Fig. 2 two segments are double-hit by two pairs of lines; the blue lines are l-hitters and the red lines are r-hitters. When we pair l_1 to l_3 and pair l_2 to l_4, the two segments are still double-hit, because this swap moves the l-hitter further left and the r-hitter further right. A sequence of such swaps moves all l-hitters to the left of all r-hitters.   □

In the full paper, we give details of an algorithm for HRVL. The algorithm maximizes the number of 3-intersections and "balances" the remaining lines between the left and right sides as much as possible. In the algorithm, we test the criticality of a line: given the previous choices, if a *critical* line is not used as a 3-hitter, there is no way to extend the previous choices to a maximum matching.

We now argue that the left-right-balanced approach gives lines that obey Lemma 2. Let S be the solution given by our HRVL algorithm, and let S' be an optimal solution with the maximum set of 3-hitters. We know that S and S' have the same number of 3-hitters. Let D and D' denote the lines left behind (not 3-hitters) in S and S' respectively. We order lines in D and D' from left to right. Let k be $\lfloor \frac{|D|}{2} \rfloor$. Thus, there are at most k pairs of double-hitters in S and S'. Let lh_i (resp., lh'_i) be the ith line of D (resp., D').

Given a solution P and a line l, let $E(l, P)$ denote the number of segments on the left side of l not hit by 3-hitters in P. A line having more segments on its right side is more likely to be an l-hitter. We will show that line lh_i is at least as capable of being an l-hitter as is line lh'_i.

$$E(lh_i, S) \le E(lh'_i, S'), \quad i = 1, 2, .., k \tag{1}$$

We split the proof of (1) into two lemmas; proofs appear in the full paper.

Lemma 3. lh_i *cannot be on the right side of* lh'_i, $i = 1, 2, .., k$.

An immediate result from this lemma is

$$E(lh_i, S') \le E(lh'_i, S'). \tag{2}$$

Given a solution P and a line l, let $C(l, P)$ denote the number of segments on the left side of l that have been 3-hit in P. Let $N(l)$ be the total number of segments on the left of line l. The following lemma shows that the segments that S leaves to be used as 2-hitters are the segments that are easier to double-hit.

Lemma 4. $C(lh_i, S) \ge C(lh_i, S')$, $i = 1, 2, .., k$.

Therefore we obtain

$$\begin{aligned}
E(lh_i, S) &= N(lh_i) - C(lh_i, S) \\
&\le N(lh_i) - C(lh_i, S') = E(lh_i, S') \le E(lh'_i, S').
\end{aligned}$$

5   Hitting Lines and Segments

5.1   Hardness

Theorem 4. *Hitting set for horizontal unit segments and vertical lines is NP-complete.*

Proof. The reduction is from 3SAT. See Fig. 3.

Each variable is represented by a collinear connected set of horizontal unit segments, and each clause is represented by a red vertical line that intersects appropriate pairs of horizontal variable segments (if that variable occurs in a clause) or just single segments (in case a variable does not occur in a clause). Setting appropriate parities for the literals in a clause is achieved by appropriate horizontal shifting of the segments, as shown in the figure. This results in a construction in which the only place where three of the elements (segments or lines) can be hit involves a vertical line representing a clause, corresponding

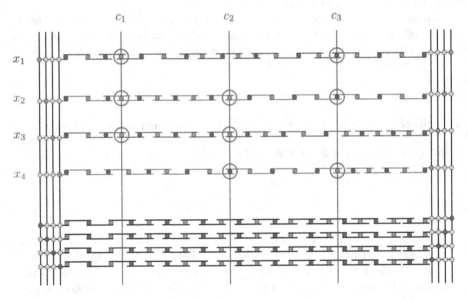

Fig. 3. A set of horizontal unit segments and vertical lines that represents the 3SAT instance $I = (x_1 \lor x_2 \lor x_3) \land (x_2 \lor x_3 \lor \overline{x_4}) \land (\overline{x_1} \lor \overline{x_2} \lor x_4)$. For better visibility, collinear segments are slightly shifted vertically, with red and green points indicating overlapping segments. In an optimal hitting set, the point covering a *labeled* horizontal segment induces a truth value for the corresponding variable: selecting one of its grey points (e.g., in the indicated green manner) assigns a value of "true"; selecting a red point, a value of "false". Overall, truth assignments for each variable correspond to a set of green or red points, respectively. Literals occurring in clauses are indicated by magenta circles; these are the only places where a point can hit three segments or lines at once (Color figure online).

to literals occurring in the respective clauses. (These are indicated by magenta circles in the figure). The N elements excluding the red vertical lines associated with clauses are called variable components.

We show that any feasible hitting set with exactly $N/2$ points induces a truth assignment and vice versa. There is no point that hits more than two of the variable components at once. Therefore, stabbing all N of them requires at least $N/2$ points, and any solution consisting of exactly $N/2$ points must hit each variable component exactly once. Consequently, the black vertical lines and the black collinear sets of connected horizontal segments in Fig. 3 may be reordered so that, without loss of generality, a solution with exactly $N/2$ points does not pick any of the gray points. Eliminating the gray points results in a natural partition of the instance into point-disjoint even-cardinality loops of variable components for each variable, where the points in each loop alternate between red and green. Thus any solution of size $N/2$ hitting the variable components must select all red or all green points from each variable's loop, corresponding to a truth assignment. We get an overall feasible hitting set if and only if the points also stab the vertical clause lines, corresponding to a satisfying truth assignment.   □

After appropriate vertical scaling, we can replace the vertical lines by vertical unit segments, immediately giving the following corollary.

Corollary 1. *Deciding if there exists a set of k points in the plane that hit a given set S of unit-length axis-parallel segments is NP-complete.*

We show in the full paper, using a reduction from MAX-2SAT(3), that the hitting set problem is APX-hard for vertical lines and horizontal segments.

5.2   Approximation

We give a 5/3-approximation for hitting a set V of vertical lines and a set H of horizontal segments. We start by looking at the lower bounds: $v = |V|$ is the number of vertical lines. It is a lower bound. Let h be the lower bound on hitting horizontal segments only. We can compute h exactly; it is the minimum number of hit points for the horizontal segments (computed on each horizontal line). At any stage of the algorithm, we let h and v be the current values of these lower bounds for hitting the current (remaining unhit) sets H and V.

In stage 1, we place two kinds of points:

(a) We place hitting points on vertical lines that reduce h (and v) by one. These points are "maximally productive" since no single hitting point can do more than to reduce h and v each by one. As vertical lines are hit, we remove them from V. Similarly, as horizontal segments are hit, we remove them from H.

(b) Look for pairs (if any) of points, on the same horizontal line and on two vertical lines (from among the current set V), that decrease h by one.

Let k_1 and k_2 be the number of type(a) and type(b) points placed in this stage, respectively. Therefore, for the remaining instance, the lower bound h decreases by $k_1 + k_2/2$, and v decreases by $k_1 + k_2$.

In stage 2, we now have a set of vertical lines V and horizontal segments H such that no single point at the intersection of a vertical line and a horizontal segment (or segments) reduces h, and no pair of points on two distinct vertical lines reduces h.

Lemma 5. *For such sets V and H as in stage 2, an optimal hitting set has size at least $v + h$, where $v = |V|$ and h is the minimum number of points to hit H.*

Proof. The hit points we place on V (one per line) might conceivably decrease h. We claim that this cannot happen. Assume to the contrary that it happens. Let $\{q_1, q_2, \ldots, q_K\}$ be a minimum-cardinality set such that each of them is on some line of V from left to right and h is decreased by 1 after placing the set. Since the set is minimum, the points in it should be on a horizontal line L.

Since we have found all productive points and pairs of points in stage 1, K should be at least 3. Consider the hit point q_2. The segments on L that are not hit by q_2 are either completely left or right of q_2; let H_l and H_r be the corresponding sets. Points to the left of q_2 do not hit H_r, and points to the right of q_2 to not hit H_l. If adding q_1 decreases H, that means q_1 and q_2 is a productive pair, which should be found in stage 1; otherwise this means that the point q_1 is unnecessary, contradicting the minimality of K.

Theorem 5. *There is a polynomial-time 5/3-approximation algorithm for geometric hitting set for a set of vertical lines and horizontal segments.*

Proof. The total number of points selected by our algorithm is $k_1 + k_2$ from the first stage and $h - k_1 - k_2/2 + v - k_1 - k_2$ from the second stage. By Lemma 5, the points chosen in stage 2 is a lower bound on the cost of an optimal solution:

$$h - k_1 - k_2/2 + v - k_1 - k_2 \leq OPT. \tag{3}$$

We also have $h \leq OPT$ and $v \leq OPT$. There are two cases.

(i) $k_1 + k_2 \leq 2/3 \cdot OPT$: In this case we select at most $2/3 \cdot OPT$ points in Stages 1, and we use (3) to bound the number of points selected in Stage 2. We conclude that our algorithm selects at most $5/3 \cdot OPT$ points.

(ii) $k_1 + k_2 > 2/3 \cdot OPT$: The total number of points selected by our algorithm is $h - k_1 - k_2/2 + v \leq 2 \cdot OPT - (k_1 + k_2/2)$. Since $k_1 + k_2/2 \geq k_1/2 + k_2/2 > 1/3 \cdot OPT$, we obtain a 5/3-approximation in this case as well. $\square$

Theorem 6. *There is a polynomial-time 5/3-approximation algorithm for geometric hitting set for a set of vertical (downward) rays and horizontal segments.*

Proof. The 2-stage approximation algorithm described above works for this case as well. The key observation is that among any set of collinear downward rays, we may remove all but the one with the lowest apex from the instance. Therefore after Stage 1, the hitting points we place on the rays not yet hit will not decrease h. The argument is analogous to that in Lemma 5.

6   Hitting Pairs of Segments

We consider now the hitting set problem for inputs that are *unions* of two segments, one horizontal and one vertical. While we are motivated by pairs (and larger sets) of segments that form paths, our methods apply to general pairs of segments, which might meet to form an "L" shape, a "+", or a "T" shape, or they may be disjoint. This hitting set problem is NP-hard, since it generalizes the case of horizontal and vertical segments.

Theorem 7. *For objects that are unions of a horizontal and a vertical segment, the hitting set problem has a polynomial-time 4-approximation.*

Proof Sketch. For ease of discussion, we call the union of two segments an "L." We use a method similar to those used in [9,20]. Solve the natural set-cover linear programming (LP) relaxation. Create two new problems: one that has only the horizontal piece of some of the Ls and another that has only the vertical pieces of the remaining Ls. Place an L into the vertical problem if the LP vertical segment has value at least $1/2$, and into the horizontal problem otherwise. Solve the two new problems in polynomial time using the combinatorial method for the 1D problem, or solving the LPs, which are totally unimodular, and thus will return integer solutions. Take all the points selected by either new problem. We prove in the full paper that these points are a 4-approximation.

The above idea naturally extends to a 4-approximation for the weighted version of the problem. For unions consisting of at most k segments drawn from r orientations, the approach yields a $(k \cdot r)$-approximation. Using similar methods and a stronger version of Theorem 5, we also have the following (see full paper):

Theorem 8. *For objects that are unions of a horizontal segment and a vertical line, the hitting set problem has a polynomial-time 10/3-approximation.*

Acknowledgment. This work is supported by the Laboratory Directed Research and Development program at Sandia National Laboratories, a multi-program laboratory managed and operated by Sandia Corporation, a wholly owned subsidiary of Lockheed Martin Corporation, for the U.S. Department of Energy's National Nuclear Security Administration under contract DE-AC04-94AL85000. J. Mitchell acknowledges support from the US-Israel Binational Science Foundation (grant 2010074) and the National Science Foundation (CCF-1018388, CCF-1526406).

References

1. Alon, N.: A non-linear lower bound for planar epsilon-nets. Discrete Comput. Geom. **47**, 235–244 (2012)
2. Aronov, B., Ezra, E., Sharir, M.: Small-size ε-nets for axis-parallel rectangles and boxes. SIAM J. Computing **39**, 3248–3282 (2010)
3. Brimkov, V.E.: Approximability issues of guarding a set of segments. Int. J. Comput. Math. **90**, 1653–1667 (2013)

4. Brimkov, V.E., Leach, A., Mastroianni, M., Wu, J.: Experimental study on approximation algorithms for guarding sets of line segments. In: Bebis, G., et al. (eds.) ISVC 2010, Part I. LNCS, vol. 6453, pp. 592–601. Springer, Heidelberg (2010)
5. Brimkov, V.E., Leach, A., Mastroianni, M., Wu, J.: Guarding a set of line segments in the plane. Theoret. Comput. Sci. **412**, 1313–1324 (2011)
6. Brimkov, V.E., Leach, A., Wu, J., Mastroianni, M.: Approximation algorithms for a geometric set cover problem. Discrete Appl. Math **160**, 1039–1052 (2012)
7. Brodén, B., Hammar, M., Nilsson, B.J.: Guarding lines and 2-link polygons is APX-hard. In: Proceedings of 13th Canadian Conference on Computational Geometry, pp. 45–48 (2001)
8. Brönnimann, H., Goodrich, M.T.: Almost optimal set covers in finite VC-dimension. Discrete Comput. Geom. **14**, 263–279 (1995)
9. Carr, R., Fujito, T., Konjevod, G., Parekh, O.: A 2 1/10-approximation algorithm for a generalization of the weighted edge-dominating set problem. In: Paterson, M. (ed.) ESA 2000. LNCS, vol. 1879, pp. 132–142. Springer, Heidelberg (2000)
10. Chvátal, V.: A greedy heuristic for the set-covering problem. Math. Oper. Res. **4**, 233–235 (1979)
11. Clarkson, K.L.: Algorithms for polytope covering and approximation. In: Dehne, F., Sack, J.-R., Santoro, N., Whitesdes, S. (eds.) Proceedings of 3rd Workshop Algorithms and Data Structures. LNCS, vol. 709, pp. 246–252. Springer, Heidelberg (1993)
12. Clarkson, K.L., Varadarajan, K.: Improved approximation algorithms for geometric set cover. Discrete Comput. Geom. **37**, 43–58 (2007)
13. Dinur, I., Steurer, D.: Analytical approach to parallel repetition. In: Proceedings of the 46th ACM Symposium on Theory of Computing, pp. 624–633 (2014)
14. Dom, M., Fellows, M.R., Rosamond, F.A., Sikdar, S.: The parameterized complexity of stabbing rectangles. Algorithmica **62**, 564–594 (2012)
15. Duh, R.-C., Fürer, M.: Approximation of k-set cover by semi-local optimization. In: Proceedings of the 29th ACM Symposium on Theory of Computing, pp. 256–264 (1997)
16. Dumitrescu, A., Jiang, M.: On the approximability of covering points by lines and related problems. Comput. Geom. **48**, 703–717 (2015)
17. Even, G., Levi, R., Rawitz, D., Schieber, B., Shahar, S.M., Sviridenko, M.: Algorithms for capacitated rectangle stabbing and lot sizing with joint set-up costs. ACM Trans. Algorithms **4**, 34:1–34:17 (2008)
18. Even, G., Rawitz, D., Shahar, S.M.: Hitting sets when the VC-dimension is small. Inf. Proc. Letters **95**, 358–362 (2005)
19. Gaur, D.R., Bhattacharya, B.: Covering points by axis parallel lines. In: Proceedings of 23rd European Workshop on Computational Geometry, pp. 42–45 (2007)
20. Gaur, D.R., Ibaraki, T., Krishnamurti, R.: Constant ratio approximation algorithms for the rectangle stabbing problem and the rectilinear partitioning problem. J. Algorithms **43**, 138–152 (2002)
21. Giannopoulos, P., Knauer, C., Rote, G., Werner, D.: Fixed-parameter tractability and lower bounds for stabbing problems. Comput. Geom. **46**, 839–860 (2013)
22. Hassin, R., Megiddo, N.: Approximation algorithms for hitting objects with straight lines. Discrete Appl. Math. **30**, 29–42 (1991)
23. Heednacram, A.: The NP-hardness of covering points with lines, paths and tours and their tractability with FPT-algorithms. Ph.D. Thesis, Griffith University (2010)
24. Hochbaum, D.S., Maas, W.: Approximation schemes for covering and packing problems in image processing and VLSI. J. ACM **32**, 130–136 (1985)

25. Joshi, A., Narayanaswamy, N.S.: Approximation algorithms for hitting triangle-free sets of line segments. In: Ravi, R., Gørtz, I.L. (eds.) SWAT 2014. LNCS, vol. 8503, pp. 357–367. Springer, Heidelberg (2014)

26. Kovaleva, S., Spieksma, F.C.: Approximation algorithms for rectangle stabbing and interval stabbing problems. SIAM J. Discrete Math. **20**, 748–768 (2006)

27. Kratsch, S., Philip, G., Ray, S.: Point line cover: the easy kernel is essentially tight. In: Proceedings of 25th ACM-SIAM Symposium on Discrete Algorithms, pp. 1596–1606 (2014)

28. Kumar, V.S.A., Arya, S., Ramesh, H.: Hardness of set cover with intersection 1. In: Welzl, E., Montanari, U., Rolim, J.D.P. (eds.) ICALP 2000. LNCS, vol. 1853, pp. 624–635. Springer, Heidelberg (2000)

29. Langerman, S., Morin, P.: Covering things with things. Discrete Comput. Geom. **33**, 717–729 (2005)

30. Megiddo, N., Tamir, A.: On the complexity of locating linear facilities in the plane. Oper. Res. Lett. **1**, 194–197 (1982)

31. Mustafa, N.H., Ray, S.: Improved results on geometric hitting set problems. Discrete Comput. Geom. **44**, 883–895 (2010)

32. O'Rourke, J.: Art Gallery Theorems and Algorithms. The International Series of Monographs on Computer Science. Oxford University Press, New York (1987)

33. Pach, J., Tardos, G.: Tight lower bounds for the size of epsilon-nets. J. Am. Math. Soc. **26**, 645–658 (2013)

34. Urrutia, J.: Art Gallery and Illumination Problems. Handbook of Computational Geometry. Elsevier, Amsterdam (1999). Chap. 22

On Independent Set on B1-EPG Graphs

M. Bougeret$^{(\boxtimes)}$, S. Bessy, D. Gonçalves, and C. Paul

LIRMM, CNRS, Université de Montpellier, Montpellier, France
`marin.bougeret@lirmm.fr`

Abstract. In this paper we consider the Maximum Independent Set problem (MIS) on B_1-EPG graphs. EPG (for Edge intersection graphs of Paths on a Grid) was introduced in [8] as the class of graphs whose vertices can be represented as simple paths on a rectangular grid so that two vertices are adjacent if and only if the corresponding paths share at least one edge of the underlying grid. The restricted class B_k-EPG denotes EPG-graphs where every path has at most k bends. The study of MIS on B_1-EPG graphs has been initiated in [6] where authors prove that MIS is NP-complete on B_1-EPG graphs, and provide a polynomial 4-approximation. In this article we study the approximability and the fixed parameter tractability of MIS on B_1-EPG. We show that there is no PTAS for MIS on B_1-EPG unless P=NP, even if there is only one shape of path, and even if each path has its vertical part or its horizontal part of length at most 3. This is optimal, as we show that if all paths have their horizontal part bounded by a constant, then MIS admits a PTAS. Finally, we show that MIS is FPT in the standard parameterization on B_1-EPG restricted to only three shapes of path, and W_1-hard on B_2-EPG. The status for general B_1-EPG (with the four shapes) is left open.

1 Introduction and Related Work

In this paper we consider the Maximum Independent Set (MIS) on B_1-EPG graphs. EPG (for Edge intersection graphs of Paths on a Grid) was introduced in [8] as the class of graphs whose vertices can be represented as simple paths on a rectangular grid so that two vertices are adjacent if and only if the corresponding paths share at least one edge of the underlying grid. More precisely, for every EPG-graph $G = (V, E)$, there exists an EPG-representation $\langle \mathcal{P}, \mathcal{G} \rangle$ where $\mathcal{P} = \{P_v, v \in V\}$ is a set of paths on the grid $\mathcal{G}$, and two paths P_u, P_v share a grid edge of $\mathcal{G}$ if and only if $\{u, v\} \in E$ (see example depicted Fig. 1). Notice that two paths can cross on a grid vertex without creating an edge. Thus, given an EPG-representation, the objective of the MIS is to find the largest set of paths that do not share a common grid edge. For any integer $k \geq 0$, the class B_k-EPG denotes EPG-graphs having an EPG-representation where every path has at most k bends. Moreover, the class X-EPG $\subseteq B_1$-EPG (with $X \subseteq \{\ulcorner, \urcorner, \llcorner, \lrcorner\}$) denotes the subset of B_1-EPG graphs where the paths can only have shapes in X (and thus B_1-EPG = $\{\ulcorner \urcorner \llcorner \lrcorner\}$-EPG).

This work was supported by the grant EGOS ANR-12-JS02-002-01.

L. Sanità and M. Skutella (Eds.): WAOA 2015, LNCS 9499, pp. 158–169, 2015.
DOI: 10.1007/978-3-319-28684-6_14

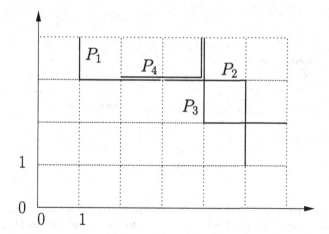

Fig. 1. B1-EPG representation of $G = (\{1,2,3,4\}, \{\{1,4\}, \{2,4\}, \{3,4\}\})$ where $ver(P_1) = [3,4]$, $ver(P_1)$ lies on column 1, $hor(P_1) = [1,3]$, $hor(P_1)$ lies on row 3, and $cor(P_1) = (1,3)$. P_1 and P_2 do not create an edge: even if $hor(P_2) = [3,5]$ and $hor(P_2)$ also lies on row 3, P_1 and P_2 do not share a common grid edge. P_2 and P_3 do not create an edge either.

For any path P with 0 or 1 bend, we define $hor(P)$ (resp. $ver(P)$) as the interval corresponding to the projection of P on the horizontal (resp. vertical) axis. Moreover, to describe the position of a path we will say that $hor(P)$ (resp. $ver(P)$) **lies** on a given row (resp. column) (see Fig. 1). Finally, we denote by $cor(P) = (x,y)$ the coordinates of the corner of P. If P has no corner, let $cor(P)$ be the coordinates of some end of P.

Related work on class inclusions. It is proved in [8] that every graph G is an EPG-graph, and that the size of the underlying grid is polynomial in the size of G. The maximum number of bends used in the representation has been improved in [10] where authors show that every graph of maximum degree Δ is in B_Δ-EPG.

Let us now consider graphs with small number of bends. Notice first that B_0-EPG graphs coincide with interval graphs. Several recent papers started the study EPG graphs with small number of bends. For example, it has been proved that B_1-EPG contains trees [8], and that B_2-EPG and B_4-EPG respectively contain outerplanar graphs and planar graphs [9]. We can also mention that the recognition of B_1-EPG graphs is NP-hard, even when only one shape of path is allowed [4].

In terms of forbidden induced subgraphs, it is also known that B_1-EPG graphs exclude induced suns S_n with $n \geq 4$, $K_{3,3}$, and $K_{3,3} - e$ [8].

There is also a close relation between EPG graphs and multiple interval graphs. A t-interval is the union of t disjoint intervals in the real line. A t-track interval is the union of t disjoint intervals on t disjoint parallel lines (called tracks), one interval on each track. Then, a t-interval graph (resp. a t-track

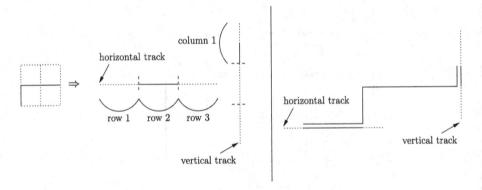

Fig. 2. (Left) B_1-EPG $\subseteq$ 2-track: one track (horizontal here) is used for the horizontal parts, and the second track (vertical here) is used for the vertical parts. (Right) 2-track $\subseteq B_3$-EPG

graph) is the intersection graph of a set of t-intervals (resp. t-track intervals). Notice that t-track interval graphs is a subclass of t-interval graphs. It is not hard to see that t-interval graphs are $B_{4(t-1)}$-EPG graphs, and that B_t-EPG graphs are $(t+1)$-interval graphs [10]. Note also that B_1-EPG graphs are 2-track graphs (use one track for the rows and one track for the columns), which in turn are B_3-EPG graphs (see Fig. 2).

Related work on MIS: Approximability. The main related article is [6] where authors show that MIS is NP-complete on B_1-EPG graphs, and provide a polynomial 4-approximation. As B_1-EPG graphs are 2-track graphs, related work on MIS on t-track graph is of interest as well. MWIS (the generalization of MIS where each vertex have an arbitrary weight) admits a $2t$-approximation on t-interval graphs [2] (see also [7]). Notice that this answers the open problem of [6] about finding approximation algorithm for maximum weighted independent set on B_1-EPG graphs. It is also proved in [2] that MIS is APX-hard on 2-track graphs, even when every vertex is represented by two intervals of length 2. We can also mention [12] that aggregates and classifies several approximation algorithms for MIS on intersection graphs by introducing several parameters. In particular the authors define the notion of k-simplicial graphs. A graph is k-simplicial if and only if there exists an order $v_1, \ldots, v_n$ of the vertices such that, for each vertex v_i, the subset of neighbors of v_i contained in $\{v_j | j > i\}$ can be partitioned into k sets $S_1, \ldots, S_k$ such that $G[S_j \bigcup \{v_i\}]$ is a clique for each $j \in \{1, \ldots, k\}$. Then they recall that MWIS is k-approximable in k-simplicial graphs. The k-simplicial graphs are related to B_1-EPG graphs. Indeed the proof of the 4-approximation of MIS on B_1-EPG graphs [6] amounts to showing that $\{\llcorner\lrcorner\}$-EPG graphs are 2-simplicial graphs, and thus that MIS has a 2-approximation on $\{\llcorner\lrcorner\}$-EPG graphs, and a 4-approximation on B_1-EPG graphs. Finally, notice that the known approximation algorithms for maximum independent set on pseudo disks intersection graphs [5] do not directly apply here, as two paths can cross on a grid vertex without creating an edge.

Related work on MIS: Fixed Parameter Tractability. In this article we only consider the decision problem $OPT \leq k$? parameterized by k. The main related result is [11] where the author proves that MIS is $W[1]$-complete on unit 2-track graphs. This immediately implies that MIS is $W[1]$-hard on B_3-EPG graphs. About positive results (*FPT* algorithm), to the best of our knowledge it seems that there is no known *FPT* algorithm on B_1-EPG or a superclass of it.

Contributions. In this article we study the approximability (in Sect. 2) and the fixed parameter tractability (in Sect. 3) of MIS on B_1-EPG. Our main results are the following. We first show that there is no PTAS for MIS on $\{\ulcorner\}$-EPG unless P=NP, even if each path has its vertical part or its horizontal part of length at most 3 (*i.e.* $\forall P, (|hor(P)| \leq 3$ or $|ver(P)| \leq 3)$). This improves the NP-hardness of [6]. Then, we show that this result cannot be improved by showing that if $\forall P, |hor(P)| \leq c$, or if $\forall P, |ver(P)| \leq c$ (where c is a constant), then MIS admits a PTAS on B_1-EPG graphs. In Sect. 3, we show that MIS is FPT on B_1-EPG restricted to only three shapes of paths, and W_1-hard on B_2-EPG. The status for general B_1-EPG (with the four shapes) is left open. Proofs marked with a $\star$ are omitted due to space constraints, and we refer the reader to [3] for the full version of this article.

2   Approximability

The objective of this section is to prove that there is no PTAS for MIS on $\{\ulcorner\}$-EPG graphs. Let us first recall that the NP-hardness proof of MIS on $\{\ulcorner\urcorner\llcorner\lrcorner\}$-EPG graphs of [6] is a reduction from MIS on planar graphs. As there is a PTAS for MIS on planar graphs, this is not a good candidate to reduce from when looking for inapproximability. Moreover, the proof of [2] showing the APX-hardness of MIS on unit 2-track graph cannot be adapted too. Indeed, this proof consists in proving that the class of unit 2-track graphs (more precisely 2-track graphs where every vertex is represented by two intervals of length 2) contains all graphs of degree at most 3 (for which MIS is APX-hard), but this class contains $K_{3,3}$ and is hence not included in B_1-EPG [10].

Our reduction is based on the classical approximation preserving reduction from MAX-3-SAT to MIS (see for example [14], Theorem 29.13).

Let us define MAX-3-SAT(3):

– Input:
  - A set of n_{var} variables $\{x_i, 1 \leq i \leq n_{var}\}$.
  - A set of m_{cl} clauses $\{C_j, 1 \leq j \leq m_{cl}\}$, where each clause is of the form $l_1^j \vee l_2^j \vee l_3^j$, where for any $1 \leq t \leq 3$, $\exists i$ such that $l_t^j = x_i$ or $l_t^j = not(x_i)$ (in both cases we say that C_j contains variable x_i).
  - Furthermore, each variable appears at most 3 times ($\forall i, |\{C_j | C_j$ contains $x_i\}| \leq 3$). Moreover, we can assume that for any i, the positive form x_i appears exactly 2 times, and the negative form appears exactly 1 time.

– Output: a truth assignment of the variables that maximizes the number of satisfied clauses

Let us define the same function f as in [14] (Theorem 29.13) that maps any instance I_{sat} of MAX-3-SAT(3) to a graph $f(I_{sat}) = G$, where $G = (V, E)$. For each clause $l_1^j \vee l_2^j \vee l_3^j$ we create a triangle $\{v_1^j, v_2^j, v_3^j\}$, and thus we have $|V| = 3m_{cl}$. For any j and t, we say that v_t^j corresponds to variable x_i (resp. to the negation of variable x_i) if and only if $l_t^j = x_i$ (resp. $l_t^j = not(x_i)$). For any $i, 1 \leq i \leq n_{var}$, we add an edge $\{v_t^j, v_{t'}^{j'}\}$ if and only if $\exists i$ such that l_t^j corresponds to x_i and $l_{t'}^{j'}$ corresponds to the negation of x_i.

Let $\mathcal{F} = \{f(I_{sat}), I_{sat}$ instance of MAX-3-SAT(3)$\}$ be the set of graphs obtained from instances of MAX-3-SAT(3).

Proposition 1 ($\star$ Folklore). *There is a strict reduction from MAX-3-SAT(3) to MIS on graphs $\mathcal{F}$.*

Proposition 2. $\mathcal{F} \subseteq \{\ulcorner, \urcorner\}$-*EPG*.

Proof. Let us draw a graph $G \in \mathcal{F}$ using only paths of the form $\{\ulcorner, \urcorner\}$. See Fig. 3 for an example. Informally, each clause j corresponds to column $3j$, and each variable x_i corresponds to line i. Let us now define more precisely the shapes of the paths.

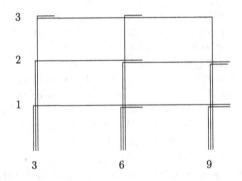

Fig. 3. Example for $C_1 = not(x_1) \vee x_2 \vee x_3$, $C_2 = x_1 \vee not(x_2) \vee x_3$, and $C_3 = x_1 \vee x_2 \vee not(x_3)$}

Let (V, E) be the vertex and edge set of G, with $|V| = n$. For each vertex $v_t^j \in V$ corresponding to a variable x_i or its negation, we define a path $P_{v_t^j}$, with $ver(P_{v_t^j}) = [0, i]$ lying on column $3j$. Notice that we already have all the vertices of V, and the edges corresponding to the $\frac{n}{3}$ triangles. It remains now to define the horizontal parts of the paths to encode the adjacencies between a variable and its negation. Let i in $\{1, \ldots, n_{var}\}$, and let $\{v_{t_1}^{j_1}, v_{t_2}^{j_2}, v_{t_3}^{j_3}\}$ be the vertices of G corresponding to variable x_i or to its negation, with $j_1 < j_2 < j_3$. There are now three cases, according to the position of the clause containing the negative form of x_i (recall that without loss of generality we suppose that $not(x_i)$ appears exactly one time).

Recall that the horizontal part of $\{v_{t_1}^{j_1}, v_{t_2}^{j_2}, v_{t_3}^{j_3}\}$ will lie one line i, and thus we just have to define the corresponding intervals. If C_{j_1} contains $not(x_i)$, then $hor(v_{t_1}^{j_1}) = [3j_1, 3j_3 + 1]$, $hor(v_{t_2}^{j_2}) = [3j_2, 3j_2 + 1]$, $hor(v_{t_1}^{j_1}) = [3j_3, 3j_3 + 1]$. If C_{j_2} contains $not(x_i)$, then $hor(v_{t_1}^{j_1}) = [3j_1, 3j_2 + 1]$, $hor(v_{t_2}^{j_2}) = [3j_2, 3j_3 + 1]$, $hor(v_{t_1}^{j_1}) = [3j_3, 3j_3 + 1]$. If C_{j_3} contains $not(x_i)$, then $hor(v_{t_1}^{j_1}) = [3j_1, 3j_1 + 1]$, $hor(v_{t_2}^{j_2}) = [3j_2, 3j_2 + 1]$, $hor(v_{t_1}^{j_1}) = [3j_1, 3j_3]$. This concludes the description of G as an $\{\ulcorner\urcorner\}$-EPG graph. □

Notice that this construction is not possible from instances of MAX-3-SAT(4), as we could have for example C_1 containing x_1, C_2 containing $not(x_1)$, C_3 containing x_1, and C_4 containing $not(x_1)$.

As MAX-3-SAT(3) remains MAXSNP-complete [13] (Theorem 13.10), we get the following corollary.

Corollary 1. *Any $(1+\epsilon)$-approximation for MIS on $\{\ulcorner\urcorner\}$-EPG implies a $(1+\epsilon)$-approximation for MAX-3-SAT(3), and thus there is no PTAS for MIS on $\{\ulcorner\urcorner\}$-EPG graphs unless P=NP.*

Our objective is now to prove that there is no PTAS for MIS, even when only one type of shape is allowed. Notice that the only case we need the $\urcorner$ shape in the previous reduction is when C_{j_3} contains $not(x_i)$.

Let $G \in \mathcal{F}$, $G = (V, E)$, with $|V| = n$ and $|E| = m$. Let us partition $E = E_1 \bigcup E_2$, where E_1 contains the n edges corresponding to the $\frac{n}{3}$ triangles, and E_2 contains the remaining edges corresponding to edges between a variable and its negation. To avoid using $\urcorner$, we will subdivide each edge of E_2 into 5 edges, introducing thus 4 new vertices for each such edge. More formally, let us define $G' = f'(G)$, $G' = (V', E')$. We start by setting $V' = V$ and $E' = E$. Moreover, for any $e = \{v_t^j, v_{t'}^{j'}\} \in E_2$ (with $j \neq j'$), we add four new vertices $w_1^e, w_2^e, w_3^e, w_4^e$ to V', and we add edges $\{v_t^j, w_1^e\}, \{\{w_i^e, w_{i+1}^e\}, 1 \leq i \leq 3\}, \{w_4^e, v_{t'}^{j'}\}$ to E'. Finally, let $\mathcal{F}' = \{f'(G), G \in \mathcal{F}\}$.

Observation 1. *Let $G = (V, E) \in \mathcal{F}$, $G' = f'(G)$. Let $m = |E|$.*

- *For any solution S of G, there is a solution S' of G' such that $|S'| \geq |S| + 2m$, and thus $Opt(G') \geq Opt(G) + 2m$*
- *For any solution S' of G', we can find in polynomial time a solution S of G such that $|S| \geq |S'| - 2m$*

From the previous observation we see that solutions of G and G' are simply shifted by an additive term of $2m$. This extra term may look too large to preserve approximability. However, the next observation shows that it only represents a constant fraction of $Opt(G)$.

Observation 2. *Let $G = (V, E) = f(I_{sat}) \in \mathcal{F}$, with $n = |V|$, $m = |E|$, n_{var} the number of variables of I_{sat}, and m_{cl} the number of clauses of I_{sat}. As G has* *maximum degree 4, $m \leq 2n$. By construction $n = 3m_{cl}$. For any 3-SAT instance it is known that $\frac{7}{8}m_{cl} \leq Opt(I_{sat})$. Finally, as $Opt(I_{sat}) = Opt(G)$, one obtains that $\frac{7}{48}m \leq Opt(G)$.*

For the next proposition, the reader might refer themselves to [3] where we recall the definition of an AP-reduction.

Proposition 3. *There is an AP-reduction from MIS on $\mathcal{F}$ to MIS on $\mathcal{F}'$.*

Proof. Point (1), (2) and (3) of the definition of an AP reduction are clearly verified. Let us prove (4) with $r = 1 + \epsilon$. Let S' be a solution of an instance G' such that $|S'| \geq \frac{Opt(G')}{(1+\epsilon)}$. According to Observation 1, we get a solution S of size at least $|S'| - 2m = \frac{Opt(G')}{1+\epsilon} - 2m \geq \frac{Opt(G)}{1+\epsilon} - 2m(\frac{\epsilon}{1+\epsilon})$. Using Observatio 2, we deduce $|S| \geq \frac{Opt(G)}{1+\epsilon}(1 - 2\epsilon\frac{48}{7})$, which concludes the proof.   □

Proposition 4. $(\star)$. $\mathcal{F}' \subseteq \{\ulcorner\}$-EPG.

We are ready to state the main inapproximability result, whose proof is now immediate.

Theorem 1. *There is no PTAS for MIS on $\{\ulcorner\}$-EPG unless P=NP, even if each path has its vertical part or its horizontal part of length at most 3 (i.e. $\forall P, (|hor(P)| \leq 3$ or $|ver(P)| \leq 3))$.*

As MIS is APX-hard on 2-track graphs [2], even when every vertex is represented by two intervals of length 2, it is natural to ask the same question here, *i.e.* to determine if Theorem 1 can be extended to paths whose vertical and horizontal intervals have constant length. Let us first notice that this restriction remains NP-hard.

Proposition 5 $(\star)$. *MIS remains NP-hard on $\{\ulcorner\}$-EPG graphs, even if all the paths have their horizontal part and their vertical part of length at most 2 (i.e. $\forall P, |hor(P)| \leq 2$ and $|ver(P)| \leq 2)$.*

As shown in the following theorem, it is not possible to improve the inapproximability, even if only one direction (but always the same) is bounded, and even if the four shapes are allowed.

Theorem 2. *MIS admits a PTAS on B_1-EPG graphs where all the paths have a horizontal part of length at most c, where c is a constant (i.e. $\forall P, |hor(P)| \leq c$). More precisely, we can find an independent set of size at least $(1 - \epsilon)|OPT|$ in $\mathcal{O}^*(n^{3c\frac{1}{\epsilon}})$ time.*

We will prove Theorem 2 using the classical Baker shifting technique [1]. We could expect that such a PTAS is already known, as MIS has been widely studied on intersection graphs (see for example the PTAS of [5] for MIS on intersection of Pseudo-Disks). However, as two paths can cross without creating an edge in the EPG model, this requires an ad-hoc adaptation of the shifting technique.

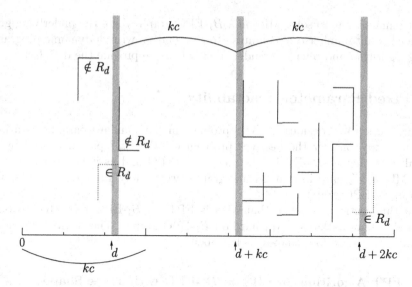

Fig. 4. R_d is the set of paths crossing one of the gray strips. We solve MIS exactly in all the white strips, *i.e.* between column $d + ack + 1$ and $d + (a + 1)ck$ with $a \geq -1$.

Proof. Let $G = (V, E)$ be a B_1-EPG graph where all the paths have an horizontal part of length at most c. Without loss of generality, let us suppose that all the paths are drawn in the positive quadrant of the plane (*i.e.* with positive coordinates), and that the underlying grid is a square of size $s \leq poly(n)$, where $n = |V|$.

Let us consider a given optimal solution OPT. Let $k \in \mathbb{N}^*$. Our goal is to get a solution of size at least $|OPT|(1 - \frac{1}{k})$.

For any integers $i \in \mathbb{N}$, let X_i be the set of paths P of G such that $[i, i+1] \subseteq hor(P)$. For any $d \in \{0, \ldots, kc - 1\}$, let $R_d = \bigcup_{a \in \mathbb{N}} X_{d+akc}$ (see Fig. 4). Let $OPT_d = OPT \cap R_d$, *i.e.* OPT_d is the set of all paths of OPT which cross one of the vertical strips, between column $d + akc$ and $d + akc + 1$ for some $a \in \mathbb{N}$.

Let $d_0 = min_d |OPT_d|$. Observe first that $|OPT_{d_0}| \leq \frac{1}{k}|OPT|$. Indeed, $\sum_{d=0}^{kc-1} |OPT_d| \leq c|OPT|$, as each vertex v^* of OPT belongs to at most c different R_l, and thus $kc|OPT_{d_0}| \leq c|OPT|$.

Thus, for each d the algorithm solves optimally the problem on $G \setminus R_d$ and gets a solution A_d. Then, it returns the largest solution among the A_d. Let us now see how we solve MIS optimally on $G' = G \setminus R_d$.

Observe that removing the vertices of R_d disconnects the graph. Indeed, paths at the left of the strip (*i.e.* P such that $hor(P) \subseteq [0, d+akc]$) cannot be connected to paths on the right of the strip (*i.e.* P such that $hor(P) \subseteq [d + akc + 1, +\infty]$). Thus, each connected component of G' correspond to a B_1-EPG graph defined on a grid with $\beta = kc$ columns. Naturally, we compute an optimal solution on G' by taking the union of optimal solutions of each connected component.

It remains now to solve MIS on a B_1-EPG graph where the underlying grid has a constant number β of columns. To that end we write a dynamic programming algorithm and refer the reader to [3] where we provide the details.   □

3   Fixed Parameter Tractability

In this section we consider the MIS problem in the standard parameterization, and thus we consider the decision problem $OPT \leq k$? parameterized by k. Recall first that as B_1-EPG $\subseteq$ 2-track $\subseteq B_3$-EPG and as it is proved in [11] that MIS is W_1-hard in unit 2-track graphs, we already know that MIS is W_1-hard in B_3-EPG graphs.

In this section we prove that MIS is FPT on B_1-EPG restricted to only three shapes of paths, and W_1-hard on B_2-EPG graphs. The status for general B_1-EPG (with the four shapes) is left open.

3.1   FPT Algorithm for MIS on B_1-EPG with Three Shapes

The principle of our FPT algorithm is to repeat the following process: locate a set of $17k^2$ paths of the instance which contains an element of an optimal solution, and then branch on these $17k^2$. If the instance is positive (i.e. has a MIS of size at most k) then the algorithm will find a solution after at most $17k^{2k}$ choices. Before detailing the algorithm, we need the following definitions and lemmas which will useful to find an element of an optimal solution.

We refer the reader to Fig. 5 for the next definitions. Notice that a purely vertical path P with $ver(P) = [a,b]$ lying on column c is considered as a path of shape ∟ with $hor(P) = [c,c]$ lying on row a. In the same way, a purely horizontal path P with $hor(P) = [a,b]$ lying on row r is considered as a path of shape ∟ with $ver(P) = [r,r]$ lying on column a.

Notice also that if G is a B_1-EPG with a path P entirely containing another path P' then $Opt(G) = Opt(G \setminus \{P\})$ for the MIS problem. Indeed, if an independent set I of G contains P then $(I \setminus \{P\}) \cup \{P'\}$ is also an independent set of G with the same size than I. Thus in every B_1-EPG instance of MIS we will consider, we assume that there is no path entirely contained in another path.

A subset G_* of paths is a **group** if and only if all paths of G_* have the same shape and the same corner (i.e. $\forall\ P_1, P_2 \in G_*\ cor(P_1) = cor(P_2)$).

Let G_* be a group of paths of shape ∟. As we can remove from the instance any path P that contains entirely another path P', we have for any $P, P' \in G_*$, $hor(P) \subseteq hor(P') \Rightarrow ver(P') \subseteq ver(P)$. We say that $P \in G_*$ **is the rightmost path of** G_* (resp. topmost) if and only if for every $P' \in G_*$ we have $hor(P') \subseteq hor(P)$ (resp. $ver(P') \subseteq ver(P)$). We also adapt the definition for the three other shapes of groups (for example for a group of shape ⌐, we define the leftmost and the downmost path).

Let S be an independent set of a B_1-EPG. Let $x \in S$ with $hor(x) = [a,b]$, lying on row r. We say that **the right side of x is free in S** (or simply that **the right side of x is free**) if there is no $y \in S$, $y \neq x$, lying on row r such that $hor(y) = [c, d]$ and $c \geq b$. We define in the same way that the left, up, and down side of x is free in S.

The **main sides** of a path are the ones given by its shape: a path of shape ∟ (resp. Γ, ¬ or ⌐) has main sides up and right (resp. down and right, down and left or up and left).

The two following lemmas provide conditions allowing us to guess in FPT time a vertex of an optimal solution.

Lemma 1. *Let r be a fixed row, and let X be a subset of paths such that every $P \in X$ has shape ∟ and $hor(P)$ lies on row r. We can construct in polynomial time a set $f_1(X) = X' \subseteq X$ with $|X'| \leq k$ verifying the following property.*

If there exists an independent set S, $|S| = k$ and a $x^* \in S \cap X$ such that the right side of x^* is free in S, then there exists S', $|S'| = k$ such that $S' \cap X' \neq \emptyset$.

Proof. Let $\{a_i : 1 \leq i \leq i_{max}\} = \{j : \exists P \in X$ with $ver(P)$ lying on column $j\}$ be the set of columns used by the vertical parts of paths of X. Let us suppose without loss of generality that $a_i > a_{i+1}$ for any i. We partition X into (maximal) groups G_i for $1 \leq i \leq i_{max}$: we define G_i as the set of all paths of X having their corner at coordinates $c_i = (a_i, r)$ (see Fig. 5). We denote by i^* the index of the group of x^* (i.e. $x^* \in G_{i^*}$). Let $X' = \{P'_1, \dots, P'_{min(i_{max},k)}\}$, where P'_i is the rightmost path of G_i. Finally, let $hor(P'_i) = [a_i, b_i]$ and $hor(x^*) = [a^*, b^*]$.

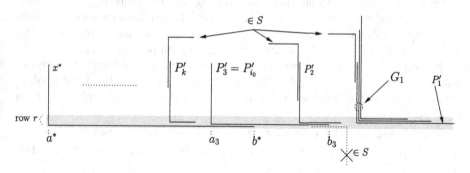

Fig. 5. Example of a group and of the case where the right of x^* is free, and $x^* \in G_{i^*}$ with $i^* > k$ in Lemma 1.

Suppose first that $i^* \leq k$. As the right side of x^* is free in S, we can remove x^* from S and chose P'_{i^*} to get another solution S' of size k.

Suppose now that $i^* > k$. Notice first that in this case by construction of the groups, we have $a^* < a_i$ for any $i \leq k$ (as we collected first paths with rightmost left endpoint). If there exists $x \in S \setminus \{x^*\}$ with $hor(x)$ lying on row r, then we have $hor(x) = [a, b]$ with $b < a^*$ (because the right side of x^* is free in OPT). So we have $b < a_i$ for any $i \leq k$ and x does not intersect paths of X'. Equivalently paths of X' can only intersect $S \setminus \{x^*\}$ by sharing vertical edges of the underlying grid. Thus as $|S| = k$, there exists $i_0 \leq k$ such that

$(S \setminus \{x^*\}) \cup \{P'_{i_0}\}$ is an independent set (*i.e.* we swap x^* and P'_{i_0} to get S'), and X' satisfies the claimed property.

$\square$

Lemma 2 ($\star$). *Let r be a fixed row, and let X be a subset of paths such that every $P \in X$ has shape $\lrcorner$ and $hor(P)$ lies on row r. We can construct in polynomial time a set $f_2(X) = X' \subseteq X$ with $|X'| \leq k$ verifying the following property.*

*If there exists an independent set S, $|S| = k$ and a $x^* \in S \cap X$ such that the right side **and** the top side of x^* are free in S, then there exists S', $|S'| = k$ such that $S' \cap X' \neq \emptyset$.*

Lemma 3. *Let X be a subset of paths such that every $P \in X$ has shape $\llcorner$. We can construct in polynomial time a set $f_3(X) = X' \subseteq X$ with $|X'| \leq 4k^2$ verifying the following property.*

If there exists an independent set S, $|S| = k$ and a $x^* \in S \cap X$ such that the two orthogonal sides of x^* are free in S, and one of this side is a main side of x^* (i.e. the free sides of x^* are right/up, right/down or up/left), then there exists S', $|S'| = k$ such that $S' \cap X' \neq \emptyset$.

Proof. Let $x^* \in S$. We first construct greedily, in polynomial time, a maximal independent set $A = \{P_1, \ldots, P_{|A|}\}$, $|A| \leq k$ (if $|A| > k$ we define X' as k arbitrary vertices of A). Let S_i^{ver} be the set of paths P in the input with shape $\llcorner$ and which intersects P_i by sharing a vertical edge of the underlying grid. Let also S_i^{hor} be defined in the same way for horizontal edges (notice that we may have $S_i^{ver} \cap S_i^{hor} \neq \emptyset$). As A is maximal, there exists i_0 such that x^* intersects P_{i_0}, and thus such that $x^* \in S_{i_0}^{ver}$ or $x^* \in S_{i_0}^{hor}$. Let us suppose first that $x^* \in S_{i_0}^{hor}$.

If the right side of x^* is free in S, then by Lemma 1 we know that we can replace x^* by a $x^{*'} \in f_1(S_{i_0}^{hor})$. Otherwise, the topside and the leftside of x^* are free in S. In this case we know by Lemma 2 that we can replace x^* by a $x^{*'} \in f_2(S_{i_0}^{hor})$.

Considering also the case $x^* \in S_{i_0}^{ver}$, we know that $x^{*'}$ can be chosen in $X' = \bigcup_{i=1}^{|A|} (f_1(S_i^{hor}) \cup f_2(S_i^{hor}) \cup f_1(S_i^{ver}) \cup f_2(S_i^{ver}))$ which of size at most $4k^2$.
$\square$

Lemma 4. *Let (G, k) be a instance of MIS on $\{\ulcorner \llcorner \lrcorner\}$-EPG graphs. We can construct in polynomial time a set X' with $|X'| \leq 17k^2$ verifying the following property.*

If there exists an independent set S, $|S| = k$, then there exists S', $|S'| = k$ such that $S' \cap X' \neq \emptyset$.

Proof. Let $x^* \in S$ (with $cor(x^*) = (a^*, b^*)$) be the top-right most path of S (i.e. for any $x \in S$ with $cor(x) = (a, b)$, either $b < b^*$ (Case 1) or $b = b^*$ and $a \leq a^*$ (Case 2). Notice that the only case where $b = b^*$ and $a = a^*$ is when x^* has shape $\lrcorner$ and there exists a unique $x \in S$ with shape $\ulcorner$, and $cor(x) = cor(x^*)$.

(Case 1) Let us first consider the case where for any $x \in S$ with $cor(x) = (a, b)$, either $b < b^*$ or ($b = b^*$ and $a < a^*$). Whatever the shape of x^*, x^* has always two orthogonal free sides in S, and one of this side is a main side (for instance, if x^* has shape ∟ then his two main directions are free, if x^* has shape ⌐ or ⌐, one of his main direction and another orthogonal direction are free). Thus, we apply three times Lemma 3 (one time for X containing all the paths of a given shape), and we construct in polynomial time a set Y of size at most $12k^2$ verifying the desired property.

(Case 2) This case is treated in the full version [3].

To summarize, if we want to ensure that we capture a vertex of a solution S' of size k, we set $X' = Y \cup Z$, leading to set of at most $17k^2$ paths.   □

We are now ready to prove the main result of this section whose proof is immediate using Lemma 4.

Theorem 3. *The question $OPT \geq k$ can be solved in time $O(k^{2k}poly(n))$ in B_1-EPG graph with three shapes of paths.*

3.2   W[1]-Hardness of MIS on B_2-EPG Graphs

Theorem 4 (⋆). *MIS is W[1]-hard, even restricted to B_2-EPG graphs.*

References

1. Baker, B.S.: Approximation algorithms for np-complete problems on planar graphs. J. ACM **41**(1), 153–180 (1994)
2. Bar-Yehuda, R., Halldórsson, M.M., Naor, J.S., Shachnai, H., Shapira, I.: Scheduling split intervals. SIAM J. Comput. **36**(1), 1–15 (2006)
3. Bougeret, M., Bessy, S., Gonçalves, D., Paul, C.: On independent set on b1-epg graphs (2015). CoRR arXiv:1510.00598
4. Cameron, K., Chaplick, S., Hoàng, C.T.: Edge intersection graphs of l-shaped paths in grids (2012). CoRR arxiv:1204.5702
5. Chan, T.M., Har-Peled, S.: Approximation algorithms for maximum independent set of pseudo-disks. Discrete Comput. Geom. **48**(2), 373–392 (2012)
6. Epstein, D., Golumbic, M.C., Morgenstern, G.: Approximation algorithms for B_1-EPG graphs. In: Dehne, F., Solis-Oba, R., Sack, J.-R. (eds.) WADS 2013. LNCS, vol. 8037, pp. 328–340. Springer, Heidelberg (2013)
7. Francis, M.C., Gonçalves, D., Ochem, P.: The maximum clique problem in multiple interval graphs. Algorithmica **71**, 1–25 (2013)
8. Golumbic, M.C., Lipshteyn, M., Stern, M.: Edge intersection graphs of single bend paths on a grid. Netw. **54**(3), 130–138 (2009)
9. Heldt, D., Knauer, K., Ueckerdt, T.: On the bend-number of planar and outerplanar graphs. In: Fernández-Baca, D. (ed.) LATIN 2012. LNCS, vol. 7256, pp. 458–469. Springer, Heidelberg (2012)
10. Heldt, D., Knauer, K., Ueckerdt, T.: Edge-intersection graphs of grid paths: the bend-number. Discrete Appl. Math. **167**, 144–162 (2014)
11. Jiang, M.: On the parameterized complexity of some optimization problems related to multiple-interval graphs. Theor. Comput. Sci. **411**, 4253–4262 (2010)
12. Kammer, F., Tholey, T.: Approximation algorithms for intersection graphs. Algorithmica **68**(2), 312–336 (2014)
13. Papadimitriou, C.H.: Computational Complexity. John Wiley and Sons Ltd., Chichester (2003)
14. Vazirani, V.V.: Approximation Algorithms. Springer Science & Business Media, Heidelberg (2001)

On the Smoothness of Paging Algorithms

Jan Reineke$^{(\boxtimes)}$ and Alejandro Salinger

Department of Computer Science, Saarland University, Saarbrücken, Germany
{reineke,salinger}@cs.uni-saarland.de

Abstract. We study the smoothness of paging algorithms. How much can the number of page faults increase due to a perturbation of the request sequence? We call a paging algorithm *smooth* if the maximal increase in page faults is proportional to the number of changes in the request sequence. We also introduce quantitative smoothness notions that measure the smoothness of an algorithm.

We derive lower and upper bounds on the smoothness of deterministic and randomized demand-paging and competitive algorithms. Among strongly-competitive deterministic algorithms LRU matches the lower bound, while FIFO matches the upper bound.

Well-known randomized algorithms like PARTITION, EQUITABLE, or MARK are shown not to be smooth. We introduce two new randomized algorithms, called SMOOTHED-LRU and LRU-RANDOM. SMOOTHED-LRU allows to sacrifice competitiveness for smoothness, where the trade-off is controlled by a parameter. LRU-RANDOM is at least as competitive as any deterministic algorithm while smoother.

1 Introduction

Due to their strong influence on system performance, paging algorithms have been studied extensively since the 1960s. Early studies were based on probabilistic request models [1–3]. In their seminal work, Sleator and Tarjan [4] introduced the notion of competitiveness, which relates the performance of an online algorithm to that of the optimal offline algorithm. By now, the competitiveness of well-known deterministic and randomized paging algorithms is well understood, and various optimal online algorithms [5,6] have been identified.

In this paper, we study the *smoothness* of paging algorithms. We seek to answer the following question: How strongly may the performance of a paging algorithm change when the sequence of memory requests is slightly perturbed? This question is relevant in various domains: Can the cache performance of an algorithm suffer significantly due to the occasional execution of interrupt handling code? Can the execution time of a safety-critical real-time application be safely and tightly bounded in the presence of interference on the cache? Can secret-dependent memory requests have a significant influence on the number of cache misses of a cryptographic protocol and thus give rise to a timing side-channel attack?

We formalize the notion of smoothness by identifying the performance of a paging algorithm with the number of page faults and the magnitude of a perturbation with the edit distance between two request sequences.

© Springer International Publishing Switzerland 2015
L. Sanità and M. Skutella (Eds.): WAOA 2015, LNCS 9499, pp. 170–182, 2015.
DOI: 10.1007/978-3-319-28684-6_15

We show that for any deterministic, demand-paging or competitive algorithm, a single additional memory request may cause $k + 1$ additional faults, where k is the size of the cache. Least-recently-used (LRU) matches this lower bound, indicating that there is no trade-off between competitiveness and smoothness for deterministic algorithms. In contrast, First-in first-out (FIFO) is shown to be least smooth among all strongly-competitive deterministic algorithms.

Randomized algorithms have been shown to be more competitive than deterministic ones. We derive lower bounds for the smoothness of randomized, demand-paging and randomized strongly-competitive algorithms that indicate that randomization might also help with smoothness. However, we show that none of the well-known randomized algorithms MARK, EQUITABLE, and PARTITION is smooth. The simple randomized algorithm that evicts one of the cached pages uniformly at random is shown to be as smooth as LRU, but not more.

We then introduce a new parameterized randomized algorithm, SMOOTHED-LRU, that allows to sacrifice competitiveness for smoothness. For some parameter values SMOOTHED-LRU is smoother than any randomized strongly-competitive algorithm can possibly be, indicating a trade-off between smoothness and competitiveness for randomized algorithms. This leaves the question whether there is a randomized algorithms that is smoother than any deterministic algorithm without sacrificing competitiveness. We answer this question in the affirmative by introducing LRU-RANDOM, a randomized version of LRU that evicts older pages with a higher probability than younger ones. We show that LRU-RANDOM is smoother than any deterministic algorithm for $k = 2$. While we conjecture that this is the case as well for general k, this remains an open problem.

The notion of smoothness we present is not meant to be an alternative to competitive analysis for the evaluation of the performance of a paging algorithm; rather, it is a complementary quantitative measure that provides guarantees about the performance of an algorithm under uncertainty of the input. In general, smoothness is useful in both testing and verification:

- In testing: if a system is smooth, then a successful test run is indicative of the system's correct behavior not only on the particular test input, but also in its neighborhood.
- In verification, systems are shown to behave correctly under some assumption on their environment. Due to incomplete environment specifications, operator errors, faulty implementations, or other causes, the environment assumption may not always hold completely. In such a case, if the system is smooth, "small" violations of the environment assumptions will, in the worst case, result in "small" deviations from correct behavior.

An example of the latter case that motivates our present work appears in safety-critical real-time systems, where static analyses are employed to derive guarantees on the worst-case execution time (WCET) of a program on a particular microarchitecture [7]. While state-of-the-art WCET analyses are able to derive fairly precise bounds on execution times, they usually only hold for the uninterrupted execution of a single program with no interference from the environment whatsoever. These assumptions are increasingly difficult to satisfy with the adoption of preemptive scheduling or even multi-core architectures, which may introduce interference on shared resources such as caches and buses. Given a smooth cache hierarchy, it is possible to separately analyze the effects of interference on the cache, e.g. due to interrupts, preemptions, or even co-running programs on other cores. Our results may thus inform the design and analysis of microarchitectures for real-time systems [8].

Interestingly, our model shows a significant difference between LRU and FIFO, two algorithms whose theoretical performance has proven difficult to separate.

Our results are summarized in Table 1. An algorithm A is (α, β, δ)-smooth, if the number of page faults $A(\sigma')$ of A on request sequence σ' is bounded by $\alpha \cdot A(\sigma) + \beta$ whenever σ can be transformed into σ' by at most δ insertions, deletions, or substitutions of individual requests. Often, our results apply to a generic value of δ. In such cases, we express the smoothness of a paging algorithm by a pair (α, β), where α and β are functions of δ, and A is $(\alpha(\delta), \beta(\delta), \delta)$-smooth for every δ. Usually, the smoothness of an algorithm depends on the size of the cache, which we denote by k. As an example, under LRU the number of faults may increase by at most $\delta(k+1)$, where δ is the number of changes in the sequence. A precise definition of these notions is given in Sect. 3.

Table 1. Upper and lower bounds on the smoothness of paging algorithms. In the table, k is the size of the cache, δ is the distance between input sequences, H_k denotes the kth harmonic number, and γ is an arbitrary constant.

Algorithm	Lower bound	Upper bound
Deterministic, demand-paging	$(1, \delta(k+1))$	∞
Det. c-competitive with additive constant β	$(1, \delta(k+1))$	$(c, 2\delta c + \beta)$
Deterministic, strongly-competitive	$(1, \delta(k+1))$	$(k, 2\delta k)$
Optimal offline	$(1, 2\delta)$	$(1, 2\delta)$
LRU	$(1, \delta(k+1))$	$(1, \delta(k+1))$
FWF	$(1, 2\delta k)$	$(1, 2\delta k)$
FIFO	$(k, \gamma, 1)$	$(k, 2\delta k)$
Randomized, demand-paging	$(1, H_k + \frac{1}{k}, 1)$	∞
Randomized, strongly-competitive	$(1, \delta H_k)$	$(H_k, 2\delta H_k)$
EQUITABLE, PARTITION	$(1 + \epsilon, \gamma, 1)$	$(H_k, 2\delta H_k)$
MARK	$(\Omega(H_k), \gamma, 1)$	$(2H_k - 1, \delta(4H_k - 2))$
RANDOM	$(1, \delta(k+1))$	$(1, \delta(k+1))$
Evict-On-Access	$(1, \delta(1 + \frac{k}{2k-1}))$	$(1, \delta(1 + \frac{k}{2k-1}))$
SMOOTHED-LRU$_{k,i}$	$(1, \delta(\frac{k+i}{2i+1} + 1))$	$(1, \delta(\frac{k+i}{2i+1} + 1))$

2   Related Work

2.1   Notions of Smoothness

Robust control is a branch of control theory that explicitly deals with uncertainty in its approach to controller design. Informally, a controller designed for a particular set of parameters is said to be robust if it would also work well under a slightly different set of assumptions. In computer science, the focus has long been on the binary property of correctness, as well as on average- and worst-case performance. Lately, however, various notions of smoothness have received increasing attention: Chaudhuri et al. [9] develop analysis techniques to determine whether a given program computes a *Lipschitz-continuous* function. Lipschitz continuity is a special case of our notion of smoothness. Continuity is also strongly related to differential privacy [10], where the result of a query may not depend strongly on the information about any particular individual. Differential privacy proofs with respect to cache side channels [11] may be achievable in a compositional manner for caches employing smooth paging algorithms.

Doyen et al. [12] consider the robustness of sequential circuits. They determine how long into the future a single disturbance in the inputs of a sequential circuit may affect the circuit's outputs. Much earlier, but in a similar vein, Kleene [13], Perles, Rabin, Shamir [14], and Liu [15] developed the theory of definite events and definite automata. The outputs of a definite automaton are determined by a fixed-length suffix of its inputs. Definiteness is a sufficient condition for smoothness.

The work of Reineke and Grund [16] is closest to ours: they study the maximal difference in the number of page faults on the *same* request sequence starting from two different initial states for various deterministic paging algorithms. In contrast, here, we study the effect of differences in the request sequences on the number of faults. Also, in addition to only studying particular deterministic algorithms as in [16], in this paper we determine smoothness properties that apply to classes of algorithms, such as all demand-paging or strongly-competitive ones, as well as to randomized algorithms. One motivation to consider randomized algorithms in this work are recent efforts to employ randomized caches in the context of hard real-time systems [17].

2.2   The Paging Problem

Paging models a two-level memory system with a small fast memory known as cache, and a large but slow memory, usually referred to simply as memory. During a program's execution, data is transferred between the cache and memory in units of data known as pages. The size of the cache in pages is usually referred to as k. The size of the memory can be assumed to be infinite. The input to the paging problem is a sequence of page requests which must be made available in the cache as they arrive. When a request for a page arrives and this page is already in the cache, then no action is required. This is known as a *hit*. Otherwise, the page must be brought from memory to the cache, possibly requiring the eviction of another page from the cache. This is known as a *page fault* or *miss*.

A paging algorithm must decide which pages to keep in the cache in order to minimize the number of faults.

A paging algorithm is said to be *demand paging* if it only evicts a page from the cache upon a fault with a full cache. Any non-demand paging algorithm can be made to be demand paging without sacrificing performance [18].

In general, paging algorithms must make decisions as requests arrive, with no knowledge of future requests. That is, paging is an online problem. The most prevalent way to analyze online algorithms is competitive analysis [4]. In this framework, the performance of an online algorithm is measured against an algorithm with full knowledge of the input sequence, known as optimal offline or OPT. We denote by $A(\sigma)$ the number of misses of an algorithm when processing the request sequence σ. A paging algorithm A is said to be c-competitive if for all sequences σ, $A(\sigma) \leq c \cdot \text{OPT}(\sigma) + \beta$, where β is a constant independent of σ. The *competitive ratio* of an algorithm is the infimum over all possible values of c satisfying the inequality above. An algorithm is called competitive if it has a constant competitive ratio and *strongly competitive* if its competitive ratio is the best possible [5].

Traditional paging algorithm are Least-recently-used (LRU)—evict the page in the cache that has been requested least recently— and First-in first-out (FIFO)—evict the page in the cache that was brought into cache the earliest. Another simple algorithm often considered is Flush-when-full (FWF)— empty the cache if the cache is full and a fault occurs. These algorithms are k-competitive, which is the best ratio that can be achieved for deterministic online algorithms [4]. An optimal offline algorithm for paging is Furthest-in-the-future, also known as Longest-forward-distance and Belady's algorithm [1]. This algorithm evicts the page in the cache that will be requested at the latest time in the future.

A competitive ratio less than k can be achieved by the use of randomization. Important randomized paging algorithms are RANDOM—evict a page chosen uniformly at random— and MARK [19]—mark a page when it is unmarked and requested, and upon a fault evict a page chosen uniformly at random among unmarked pages (unmarking all pages first if no unmarked pages remain). RANDOM achieves a competitive ratio of k, while MARK's competitive ratio is $2H_k - 1$, where $H_k = \sum_{i=1}^{k} \frac{1}{i}$ is the kth harmonic number. The strongly-competitive algorithms PARTITION [5] and EQUITABLE [6] achieve the optimal ratio of H_k.

3 Smoothness of Paging Algorithms

We now formalize the notion of smoothness of paging algorithms. We are interested in answering the following question: How does the number of misses of a paging algorithm vary as its inputs vary? We quantify the similarity of two request sequences by their edit distance:

Definition 1 (Distance). *Let $\sigma = x_1, \ldots, x_n$ and $\sigma' = x'_1, \ldots, x'_m$ be two request sequences. Then we denote by $\Delta(\sigma, \sigma')$ their edit distance, defined as the minimum number of substitutions, insertions, or deletions to transform σ into σ'.*

This is also referred to as the *Levenshtein distance*. Based on this notion of distance we define (α, β, δ)-smoothness:

Definition 2 $((\alpha, \beta, \delta)$**-Smoothness).** *Given a paging algorithm A, we say that A is (α, β, δ)-smooth, if for all pairs of sequences σ, σ' with $\Delta(\sigma, \sigma') \leq \delta$,*

$$A(\sigma') \leq \alpha \cdot A(\sigma) + \beta$$

For randomized algorithms, $A(\sigma)$ denotes the algorithm's *expected* number of faults when serving σ.

An algorithm that is (α, β, δ)-smooth may also be $(\alpha', \beta', \delta)$-smooth for $\alpha' > \alpha$ and $\beta' < \beta$. As the multiplicative factor α dominates the additive constant β in the long run, when analyzing the smoothness of an algorithm, we first look for the minimal α such that the algorithm is (α, β, δ)-smooth for any β.

We say that an algorithm is *smooth* if it is $(1, \beta, 1)$-smooth for some β. In this case, the maximal increase in the number of page faults is proportional to the number of changes in the request sequence. This is called *Lipschitz continuity* in mathematical analysis. For smooth algorithms, we also analyze the Lipschitz constant, i.e., the additive part β in detail, otherwise we concentrate the analysis on the multiplicative factor α.

We use the above notation when referring to a specific distance δ. For a generic value of δ we omit this parameter and express the smoothness of a paging algorithm with a pair (α, β), where both α and β are functions of δ.

Definition 3 $((\alpha, \beta)$**-Smoothness).** *Given a paging algorithm A, we say that A is (α, β)-smooth, if for all pairs of sequences σ, σ',*

$$A(\sigma') \leq \alpha(\delta) \cdot A(\sigma) + \beta(\delta),$$

where α and β are functions, and $\delta = \Delta(\sigma, \sigma')$.

Often, it is enough to determine the effects of one change in the inputs to characterize the smoothness of an algorithm A.

Lemma 1. *If A is $(\alpha, \beta, 1)$-smooth, then A is $(\alpha^\delta, \beta \sum_{i=0}^{\delta-1} \alpha^i)$-smooth.*

All proofs can be found in the full version of this paper [20].

Corollary 1. *If A is $(1, \beta, 1)$-smooth, then A is $(1, \delta\beta)$-smooth.*

4 Smoothness of Deterministic Paging Algorithms

4.1 Bounds on the Smoothness of Deterministic Paging Algorithms

Before considering particular deterministic online algorithms, we determine upper and lower bounds for several important classes of algorithms. Many natural algorithms are demand paging.

Theorem 1 (Lower Bound for Deterministic, Demand-Paging Algorithms). *No deterministic, demand-paging algorithm is $(1, \delta(k+1-\epsilon))$-smooth for any $\epsilon > 0$.*

The idea of the proof is to first construct a sequence of length $k + \delta(k + 1)$ containing $k+1$ distinct pages, such that A faults on every request. This sequence can then be transformed into a sequence containing only k distinct pages by removing all requests to the page that occurs least often, which reduces the overall number of misses to k, while requiring at most δ changes. While most algorithms are demand paging, it is not a necessary condition for an algorithm to be competitive, as demonstrated by FWF. However, we obtain the same lower bound for competitive algorithms as for demand-paging ones.

Theorem 2 (Lower Bound for Deterministic, Competitive Paging Algorithms). *No deterministic, competitive paging algorithm is $(1, \delta(k+1-\epsilon))$-smooth for any $\epsilon > 0$.*

By contraposition of Corollary 1, the two previous theorems show that no deterministic, demand-paging or competitive algorithm is $(1, k + 1 - \epsilon, 1)$-smooth for any $\epsilon > 0$.

Intuitively, the optimal offline algorithm should be very smooth, and this is indeed the case as we show next:

Theorem 3 (Smoothness of OPT). *OPT is $(1, 2\delta)$-smooth. This is tight.*

With Theorem 3 it is easy to show the following upper bound on the smoothness of any competitive algorithm:

Theorem 4 (Smoothness of Competitive Algorithms). *Let A be any paging algorithm such that for all sequences σ, $A(\sigma) \leq c \cdot \text{OPT}(\sigma) + \beta$. Then A is $(c, 2\delta c + \beta)$-smooth.*

Note that the above theorem applies to both deterministic and randomized algorithms. Given that every competitive algorithm is (α, β)-smooth for some α and β, the natural question to ask is whether the converse also holds. Below, we answer this question in the affirmative for deterministic bounded-memory, demand-paging algorithms. By *bounded memory* we mean algorithms that, in addition to the contents of their fast memory, only have a finite amount of additional state. For a more formal definition consult [18, page 93]. Paging algorithms implemented in hardware caches are bounded memory.

Theorem 5 (Competitiveness of Smooth Algorithms). *If algorithm A is deterministic bounded-memory, demand-paging, and (α, β)-smooth for some α and β, then A is also competitive.*

4.2   Smoothness of Particular Deterministic Algorithms

Now let us turn to the analysis of three well-known deterministic algorithms: LRU, FWF, and FIFO. We show that both LRU and FWF are smooth. On the other hand, FIFO is not smooth, as a single change in the request sequence may increase the number of misses by a factor of k.

Theorem 6 (Smoothness of Least-Recently-Used).
LRU *is* $(1, \delta(k + 1))$*-smooth. This is tight.*

So LRU matches the lower bound for both demand-paging and competitive paging algorithms. We now show that FWF is also smooth, with a factor that is almost twice that of LRU. The smoothness of FWF follows from the fact that it always misses k times per phase, and the number of phases can only change marginally when perturbing a sequence.

Theorem 7 (Smoothness of Flush-When-Full).
FWF *is* $(1, 2\delta k)$*-smooth. This is tight.*

We now show that FIFO is not smooth. In fact, we show that with only a single difference in the sequences, the number of misses of FIFO can be k times higher than the number of misses in the original sequence. On the other hand, since FIFO is strongly competitive, the multiplicative factor k is also an upper bound for FIFO's smoothness.

Theorem 8 (Smoothness of First-in First-Out).
FIFO *is* $(k, 2\delta k)$*-smooth.* FIFO *is not* $(k - \epsilon, \gamma, 1)$*-smooth for any* $\epsilon > 0$ *and* γ.

FIFO matches the upper bound for strongly-competitive deterministic paging algorithms. With the result for LRU, this demonstrates that the upper and lower bounds for the smoothness of strongly-competitive algorithms are tight.

5   Smoothness of Randomized Paging Algorithms

5.1   Bounds on the Smoothness of Randomized Paging Algorithms

Similarly to deterministic algorithms, we can show a lower bound on the smoothness of any randomized demand-paging algorithm. The proof is strongly inspired by the proof of a lower bound for the competitiveness of randomized algorithms by Fiat et al. [19]. The high-level idea is to construct a sequence using $k + 1$ distinct pages on which any randomized algorithm faults at least $k + H_k + \frac{1}{k}$ times that can be converted into a sequence containing only k distinct pages by deleting a single request. Notice that the lower bound only applies to $\delta = 1$ and so additional disturbances might have a smaller effect than the first one.

Theorem 9 (Lower Bound for Randomized, Demand-Paging Algorithms). *No randomized, demand-paging algorithm is* $(1, H_k + \frac{1}{k} - \epsilon, 1)$*-smooth for any* $\epsilon > 0$.

For strongly-competitive randomized algorithms we can show a similar statement using a similar yet more complex construction:

Theorem 10 (Lower Bound for Strongly-Competitive Randomized Paging Algorithms). *No strongly-competitive, randomized paging algorithm is* $(1, \delta(H_k - \epsilon))$*-smooth for any* $\epsilon > 0$.

In contrast to the deterministic case, this lower bound only applies to strongly-competitive algorithms, as opposed to simply competitive. So with randomization there might be a trade-off between competitiveness and smoothness. There might be competitive algorithms that are smoother than all strongly-competitive ones.

5.2   Smoothness of Particular Randomized Algorithms

Two known strongly-competitive randomized paging algorithms are PARTITION [5] and EQUITABLE [6]. We show that neither of the two algorithms is smooth.

Theorem 11 (Smoothness of Partition and Equitable). *For any cache size $k \geq 2$, there is an $\epsilon > 0$, such that neither PARTITION nor EQUITABLE is $(1+\epsilon, \gamma, 1)$-smooth for any γ. Also, PARTITION and EQUITABLE are $(H_k, 2\delta H_k)$-smooth.*

The lower bound in the theorem above is not tight, but it shows that neither of the two algorithms matches the lower bound from Theorem 10. This leaves open the question whether the lower bound from Theorem 10 is tight.

MARK [19] is a simpler randomized algorithm that is $(2H_k - 1)$-competitive. We show that it is not smooth either.

Theorem 12 (Smoothness of Mark). *Let $\alpha = \max_{1 < \ell \leq k} \left\{ \frac{\ell(1+H_k-H_\ell)}{\ell-1+H_k-H_{\ell-1}} \right\} = \Omega(H_k)$, where k is the cache size. MARK is not $(\alpha - \epsilon, \gamma, 1)$-smooth for any $\epsilon > 0$ and any γ. Also, MARK is $(2H_k - 1, \delta(4H_k - 2))$-smooth.*

We conjecture that the lower bound for MARK is tight, i.e., that MARK is (α, β)-smooth for α as defined in Theorem 12 and some β.

We now prove that RANDOM achieves the same bounds for smoothness as LRU and the best possible for any deterministic, demand-paging or competitive algorithm. For simplicity, we prove the theorem for a non-demand-paging definition of RANDOM in which each page gets evicted upon a miss with probability $1/k$ even if the cache is not yet full. Intuitively, the additive term $k + 1$ in the smoothness of RANDOM is explained by the fact that a single difference between two sequences can make the caches of both executions differ by one page p. Since RANDOM evicts a page with probability $1/k$, the expected number of faults until p is evicted is k.

Theorem 13 (Smoothness of Random).
RANDOM *is $(1, \delta(k + 1))$-smooth. This is tight.*

5.3   Trading Competitiveness for Smoothness

We have seen that none of the well-known randomized algorithms are particularly smooth. RANDOM is the only known randomized algorithm that is $(1, \delta c)$-smooth for some c. However, it is neither smoother nor more competitive than LRU, the smoothest deterministic algorithm. In this section we show that greater smoothness can be achieved at the expense of competitiveness. First, as an extreme example of this, we show that Evict-on-access (EOA) [17]—the policy that evicts each page with a probability of $\frac{1}{k}$ upon *every* request, i.e., not only on faults but also on hits—beats the lower bounds of Theorems 9 and 10 and is strictly smoother than OPT. This policy is non-demand paging and it is obviously not competitive. We then introduce SMOOTHED-LRU, a parameterized randomized algorithm that trades competitiveness for smoothness.

Theorem 14 (Smoothness of EOA).
EOA *is* $(1, \delta(1 + \frac{k}{2k-1}))$-*smooth. This is tight.*

Smoothed-LRU. We now describe SMOOTHED-LRU. The main idea of this algorithm is to smooth out the transition from the hit to the miss case.

The following notion of *age* is convenient in the analysis of LRU: The age of page p is the number of distinct pages that have been requested since the previous request to p. LRU faults if and only if the requested page's age is greater than or equal to k, the size of the cache. An additional request may increase the ages of k cached pages by one. At the next request to each of these pages, the page's age may thus increase from $k - 1$ to k, and turn the request from a hit into a miss, resulting in k additional misses.

By construction, under SMOOTHED-LRU, the hit probability of a request decreases only gradually with increasing age. The speed of the transition from definite hit to definite miss is controlled by a parameter i, with $0 \leq i < k$. Under SMOOTHED-LRU, the hit probability $P(hit_{\text{SMOOTHED-LRU}_{k,i}}(a))$ of a request to a page with age a is:

$$P(hit_{\text{SMOOTHED-LRU}_{k,i}}(a)) = \begin{cases} 1 & : a < k - i \\ \frac{k+i-a}{2i+1} & : k - i \leq a < k + i \\ 0 & : a \geq k + i \end{cases} \quad (1)$$

where k is the size of the cache. Figure 1 illustrates this graphically in relation to LRU for cache size $k = 8$ and $i = 4$.

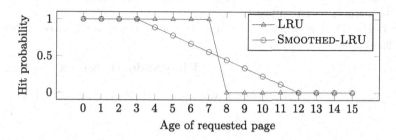

Fig. 1. Hit probabilities of LRU and SMOOTHED-LRU in terms of the age of the requested page

Theorem 15 (Smoothness of Smoothed-LRU).
SMOOTHED-LRU$_{k,i}$ *is* $(1, \delta(\frac{k+i}{2i+1} + 1))$-*smooth. This is tight.*

For $i = 0$, SMOOTHED-LRU is identical to LRU and $(1, \delta(k+1))$-smooth. At the other extreme, for $i = k - 1$, SMOOTHED-LRU is $(1, 2\delta)$-smooth, like the optimal offline algorithm. However, for larger i, SMOOTHED-LRU is less competitive than LRU:

Lemma 2 (Competitiveness of Smoothed-LRU). *For any sequence σ and $l \leq k - i$,*

$$\text{SMOOTHED-LRU}_{k,i}(\sigma) \leq \frac{k - i}{k - i - l + 1} \cdot \text{OPT}_l(\sigma) + l,$$

where $\text{OPT}_l(\sigma)$ *denotes the number of faults of the optimal offline algorithm processing σ on a fast memory of size l. For $l > k - i$ and any α and β there is a sequence σ, such that* SMOOTHED-LRU$_{k,i}(\sigma) > \alpha \cdot \text{OPT}_l(\sigma) + \beta$.

So far we have analyzed SMOOTHED-LRU based on the hit probabilities given in (1). We have yet to show that a randomized algorithm satisfying (1) can be realized. In the full version of the paper [20], we construct a probability distribution on the set of all deterministic algorithms using a fast memory of size k that satisfies (1). This is commonly referred to as a *mixed strategy*.

5.4   A Competitive and Smooth Randomized Paging Algorithm: LRU-Random

In this section we introduce and analyze LRU-RANDOM, a competitive randomized algorithm that is smoother than any competitive deterministic algorithm. LRU-RANDOM orders the pages in the fast memory by their recency of use; like LRU. Upon a miss, LRU-RANDOM evicts older pages with a higher probability than younger pages. More precisely, the ith oldest page in the cache is evicted with probability $\frac{1}{i \cdot H_k}$. By construction the eviction probabilities sum up to 1: $\sum_{i=1}^{k} \frac{1}{i \cdot H_k} = \frac{1}{H_k} \cdot \sum_{i=1}^{k} \frac{1}{i} = 1$. LRU-RANDOM is *not* demand paging: if the cache is not yet entirely filled, it may still evict cached pages according to the probabilities mentioned above.

LRU-RANDOM is at least as competitive as strongly-competitive deterministic algorithms:

Theorem 16 (Competitiveness of LRU-Random). *For any sequence σ,*

$$\text{LRU-RANDOM}(\sigma) \leq k \cdot \text{OPT}(\sigma).$$

The proof of Theorem 16 applies to an adaptive online adversary. An analysis for an oblivious adversary might yield a lower competitive ratio.

For $k = 2$, we also show that LRU-RANDOM is $(1, \delta c)$-smooth, where c is less than $k + 1$, which is the best possible among deterministic, demand-paging or competitive algorithms. Specifically, c is $1 + 11/6 = 2.8\overline{3}$. Although our proof technique does not scale beyond $k = 2$, we conjecture that this algorithm is in fact smoother than $(1, \delta(k + 1))$ for all k.

Theorem 17 (Smoothness of LRU-Random).
  Let $k = 2$. LRU-RANDOM is $(1, \frac{17}{6}\delta)$-smooth.

Conjecture 1 (Smoothness of LRU-Random).
  LRU-RANDOM is $(1, \Theta(H_k^2)\delta)$-smooth.

6   Discussion

We have determined fundamental limits on the smoothness of deterministic and randomized paging algorithms. No deterministic competitive algorithm can be smoother than $(1, \delta(k + 1))$-smooth. Under the restriction to bounded-memory algorithms, which is natural for hardware implementations of caches, smoothness implies competitiveness. LRU is strongly competitive, and it matches the lower bound for deterministic competitive algorithms. LRU is strongly competitive, and it matches the lower bound for deterministic competitive algorithms, while FIFO matches the upper bound. There is no trade-off between smoothness and competitiveness for deterministic algorithms.

In contrast, among randomized algorithms, we have identified SMOOTHED-LRU, an algorithm that is very smooth, but not competitive. In particular, it is smoother than any strongly-competitive randomized algorithm may be. The well-known randomized algorithms MARK, PARTITION, and EQUITABLE are not smooth. It is an open question, whether there is a randomized "LRU sibling"

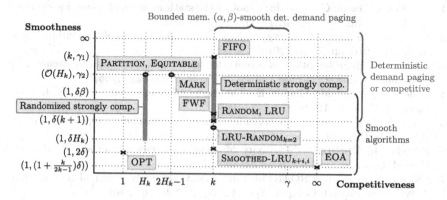

Fig. 2. Schematic view of the smoothness and competitiveness landscape. Crosses indicate tight results, whereas ellipses indicate upper bounds. Braces denote upper and lower bounds on the smoothness or competitiveness of classes of algorithms. For simplicity of exposition, γ_1 and γ_2 are left unspecified; γ can be chosen arbitrarily. More precise statements are provided in the respective theorems.

that is both strongly-competitive and $(1, \delta H_k)$-smooth. With LRU-RANDOM we introduce a randomized algorithm that is at least as competitive as any deterministic algorithm, yet provably smoother, at least for $k = 2$. Its exact smoothness remains open. Figure 2 schematically illustrates many of our results.

Acknowledgments. This work was partially supported by the DFG as part of the SFB/TR 14 AVACS.

References

1. Belady, L.A.: A study of replacement algorithms for virtual-storage computer. IBM Syst. J. **5**(2), 78–101 (1966)
2. Mattson, R.L., Gecsei, J., Slutz, D.R., Traiger, I.L.: Evaluation techniques for storage hierarchies. IBM Syst. J. **9**(2), 78–117 (1970)
3. Aho, A., Denning, P., Ullman, J.: Principles of optimal page replacement. J. ACM **18**(1), 80–93 (1971)
4. Sleator, D.D., Tarjan, R.E.: Amortized efficiency of list update and paging rules. Commun. ACM **28**(2), 202–208 (1985)
5. McGeoch, L., Sleator, D.: A strongly competitive randomized paging algorithm. Algorithmica **6**, 816–825 (1991). doi:10.1007/BF01759073
6. Achlioptas, D., Chrobak, M., Noga, J.: Competitive analysis of randomized paging algorithms. Theoret. Comput. Sci. **234**(1–2), 203–218 (2000)
7. Wilhelm, R., et al.: The worst-case execution-time problem-overview of methods and survey of tools. ACM Trans. Embed. Comput. Syst. **7**(3), 36:1–36:53 (2008)
8. Axer, P., et al.: Building timing predictable embedded systems. ACM Trans. Embed. Comput. Syst. **13**(4), 82:1–82:37 (2014)
9. Chaudhuri, S., Gulwani, S., Lublinerman, R.: Continuity and robustness of programs. Commun. ACM **55**(8), 107–115 (2012)
10. Dwork, C.: Differential privacy. In: Bugliesi, M., Preneel, B., Sassone, V., Wegener, I. (eds.) ICALP 2006. LNCS, vol. 4052, pp. 1–12. Springer, Heidelberg (2006)
11. Doychev, G., et al.: CacheAudit: a tool for the static analysis of cache side channels. ACM Trans. Inf. Syst. Secur. **18**(1), 4:1–4:32 (2015)
12. Doyen, L., Henzinger, T., Legay, A., Nickovic, D.: Robustness of sequential circuits. In: ACSD 2010, pp. 77–84 (2010)
13. Kleene, S.: Representation of events in nerve nets and finite automata. In: Automata Studies, Princeton University Press, Princeton (1956)
14. Perles, M., Rabin, M., Shamir, E.: The theory of definite automata. IEEE Trans. Electron. Comput. **12**(3), 233–243 (1963)
15. Liu, C.L.: Some memory aspects of finite automata. Technical report 411, Massachusetts Institute of Technology, May 1963
16. Reineke, J., Grund, D.: Sensitivity of cache replacement policies. ACM Trans. Embed. Comput. Syst. **12**(1s), 42:1–42:18 (2013)
17. Cazorla, F.J., et al.: PROARTIS: probabilistically analyzable real-time systems. ACM Trans. Embed. Comput. Syst. **12**(2s), 94:1–94:26 (2013)
18. Borodin, A., El-Yaniv, R.: Online Computation and Competitive Analysis. Cambridge University Press, New York (1998)
19. Fiat, A., Karp, R.M., Luby, M., McGeoch, L.A., Sleator, D.D., Young, N.E.: Competitive paging algorithms. J. Algorithms **12**(4), 685–699 (1991)
20. Reineke, J., Salinger, A.: On the smoothness of paging algorithms, October 2015. arxiv:1510.03362

Scheduling Parallel Jobs Online with Convex and Concave Parallelizability

Roozbeh Ebrahimi[1], Samuel McCauley[1,2]([✉]), and Benjamin Moseley[3]

[1] Stony Brook University, Stony Brook, NY 11794, USA
{rebrahimi,smccauley}@cs.stonybrook.edu
[2] City University of Hong Kong, Tat Chee Avenue, Kowloon, Hong Kong, China
[3] Washington University in St. Louis, St. Louis, MO 63130, USA
bmoseley@wustl.edu

Abstract. Online scheduling of parallelizable jobs has received a significant amount of attention recently. Scalable algorithms are known—that is, algorithms that are $(1+\varepsilon)$-speed $O(1)$-competitive for any fixed $\varepsilon > 0$. Previous research has focused on the case where each job's parallelizability can be expressed as a concave speedup curve. However, there are cases where a job's speedup curve can be convex. Considering convex speedup curves has received attention in the offline setting, but, to date, there are no positive results in the online model. In this work, we consider scheduling jobs with convex or concave speedup curves for the first time in the online setting. We give a new algorithm that is $(1+\varepsilon)$-speed $O(1)$-competitive. There are strong lower bounds on the competitive ratio if the algorithm is not given resource augmentation over the adversary, and thus this is essentially the best positive result one can show for this setting.

Keywords: Online scheduling · Convex and concave parallelizability · Competitive analysis

1 Introduction

Scheduling jobs online arises in numerous applications and, for this reason, there has been extensive research on the topic. See [14,20] for an overview of recent work. In general, there are n jobs that arrive over time. Each job J_i has a processing time p_i and a release date r_i. Each job J_i can only be processed after its release date and is completed once it receives p_i units of processing. In the *online* setting, the scheduler is only aware of the job after it is released.

In this study, the objective is to minimize average (total) flow time, the most popular and well-studied objective in the online setting. If a scheduler A completes job J_i at time C_i^A the *flow time* for job J_i is $F_i^A = C_i^A - r_i$. This is

---

This research was supported in part by NSF grants CNS 1408695, CCF 1439084, IIS 1247726, IIS 1251137, and CCF 1217708.

Samuel McCauley was also supported in part by Sandia National Laboratories.

L. Sanità and M. Skutella (Eds.): WAOA 2015, LNCS 9499, pp. 183–195, 2015.
DOI: 10.1007/978-3-319-28684-6_16

the amount of time the job waits to be satisfied. A scheduler A that minimizes average flow time optimizes $\sum_i F_i^A / n$, which is the average waiting time of jobs.

In the most basic setting, the jobs are to be scheduled on a single machine. In this case, it is well-known that the simple algorithm Shortest-Remaining-Processing-Time (SRPT) is optimal. Recently, there have been many results focusing on optimizing average flow time in a variety of multiple machine models; for example, see [5,6,11,13]. Much of this work has focused on the case where jobs are to be scheduled on at most one machine at any point in time. However, a significant amount of attention has also been paid to scheduling jobs that are parallelizable across m identical processors.

One of the most popular models is the arbitrary speedup curves model [7], where each job J_i has a speedup function $\Gamma(x) : m \to \mathbb{R}^+$. Here $\Gamma_i(x)$ is the rate at which job J_i is processed when given x *processing units*.

It is assumed in previous work on online scheduling that the speedup Γ_i is a non-decreasing concave function. It is interesting to note that this model captures the classic identical machine scheduling setting where a job can only be processed by at most one machine any point in time. To see this, let $\Gamma_i(x) = x$ for $x \in [0, 1]$ and 1 for $x > 1$ for all jobs J_i.

Since no algorithm can be $O(1)$-competitive for the arbitrary speedup curve model [17], previous work has focused on the resource augmentation model [16], where a job is given faster processors than the adversary. An algorithm is said to be s-speed c-competitive if the algorithm is given processors of speed s, the adversary has processors of speed 1 and the competitive ratio is c. In the speedup curve setting, this can be interpreted as $\Gamma_i(x)$ being increased by a factor s for the algorithm. An ideal algorithm is one that is $(1 + \varepsilon)$-speed $O(1)$-competitive for every fixed $\varepsilon > 0$. That is, the algorithm has a constant competitive ratio while using an arbitrarily small amount of extra resources over the adversary. Such an algorithm is called *scalable*.

It was first shown that the algorithm Round Robin, or processor sharing, is $(2 + \varepsilon)$-speed $O(1)$-competitive for any fixed $\varepsilon > 0$ [7]. This was the best positive result for roughly a decade, until a breakthrough result of Edmonds and Pruhs [9] showed an algorithm, called LAPS, is scalable. This algorithm has been extremely influential since its introduction, being shown to be the best possible algorithm in numerous scheduling environments [1,8,12].

Recently, there has also been work on determining the best competitive ratio that can be achieved if the algorithm is not given resource augmentation [15]. In this model, for concave polynomial speedup functions, there is an $O(\log P)$-competitive algorithm that is a hybrid algorithm between SRPT and Round Robin (here P is the size of the largest job); this bound is essentially tight.

Convex Speedup Functions. We offer a model of online scheduling for arbitrary speedup curves which allows speedup functions to be convex as well as concave. All of the previous work on the online arbitrary speedup curves setting only allow speedup functions to be concave. However, several works have considered convex speedup functions in the the *malleable task model*, the offline equivalent of our setting [3,4,18].

Blazewicz et al. in [4] give some examples of applications of parallel computer systems in scientific computing of highly parallelizable tasks to justify the consideration of convex speedup curves. Their examples include (1) simulation of molecular dynamics, (2) Cholesky factorization, and (3) operational oceanography.

For completeness, we summarize the molecular dynamics problem and why it experiences convex speedup. Simulation of molecular dynamics of large organic molecules (like proteins) is usually a very complicated task, because the simulation must calculate interactions between hundreds of thousands of atoms at each step. These simulations require massive amounts of space, and when an instance does not fit in main memory the cost of repeatedly writing/reading from disk dictates the time of execution. As a result, increasing the number of processors working on the problem decreases these disk accesses. This leads to a convex speedup (see Fig. 1 in [4]).

Several other studies report convex speedup curves in practice. For example, Beaumont and Guermouche in [2] report that implementing the sparse matrix factorization method of Prasanna and Musicus [19] in a real multifrontal solver results in a measured speedup function of $p^{1.15}$. They state that this superlinear speedup originates from unusual requirements on processor utilization: the algorithm generates master and slave tasks which must be scheduled on different processors.

Our Contributions. In this work, we consider scheduling jobs in the online setting with speedup curves that may be convex or concave for the first time. In our model, each job J_i is comprised of phases $\langle J_i^1, J_i^2, \cdots, J_i^{q_i} \rangle$ and each phase can have an arbitrary speedup curve. We assume that the parallelizability of phase μ of job i, J_i^μ is either a convex or concave function and each job must be assigned an integral number of processing units at each time. Note that each phase of each job can have different parallelizability, and can change from convex to concave or visa versa. Since this setting is more general than the parallel identical machines setting, there are strong lower bounds on the competitive ratio of any online algorithm.[1] Thus, we consider algorithms in the resource-augmentation model.

We assume the scheduler is online and does not know of a job until it arrives. We further assume that the scheduling algorithm does not know the size of a job nor the specific function of its speedup curve. However, we assume that the algorithm knows whether $\Gamma_{i,\mu}$ for each phase μ of job i, J_i^μ is convex or concave. That is, while that algorithm unaware of $\Gamma_{i,\mu}$ for each job J_i and phase μ, it is aware if $\Gamma_{i,\mu}$ is convex or concave. We refer to such an scheduler as a **convexity-sensitive scheduler**. We note that a convexity-sensitive scheduler has only *one bit per phase* of each job more information than non-clairvoyant LAPS algorithm in [9]. Here we show the following result.

Theorem 1. *There exists a $(1+\lambda\varepsilon)$-speed $O(\frac{1}{\lambda\varepsilon^4})$-competitive convexity-sensitive scheduling algorithm for minimizing average flow time in the arbitrary speedup*

---

[1] Specifically, [17] gives an $\Omega(\log n/m)$ lower bound.

curves setting when the speedup curves can be a concave or convex function, for any $0 < \varepsilon \leq 1/2$ and $\lambda > 7/3$.

Thus, for any fixed ε', we can achieve a $(1 + \varepsilon')$ speed, $O(1)$-competitive algorithm by appropriately choosing ε and λ in the above theorem. Given the strong lower bound on the competitive ratio for any algorithm without resource augmentation, this is essentially the best positive result that can be shown in the worst-case analysis setting.

Organization. Section 2 outlines some preliminary tools for our analysis. In Sect. 3, we present our convexity-sensitive algorithm. Section 4 provides the proof for Theorem 1.

2 Preliminaries

In our setup we consider n jobs that arrive over time. Each job J_i has a **release time** r_i and a sequence of phases $\chi_i = \langle J_i^1, J_i^2, \cdots, J_i^{q_i} \rangle$. Each phase μ of job i has a **processing time** $p_{i,\mu}$ which means that it needs this amount of processing to be completed. We assume that there are m available processing units and that when phase μ of job i is given x (*integral*) units of processing it is processed at a rate of $\Gamma_{i,\mu}(x)$. We assume that $\Gamma_{i,\mu}(0) = 0$. If the algorithm is given **resource augmentation** s, then we assume $\Gamma_{i,\mu}(x)$ is scaled by s for the algorithm (whose performance is then compared to the unscaled optimal solution).

Our algorithm assigns fractional units of processing units to jobs, which might seem to contradict the model described above. We describe how this assumption is justifiable. First, since $\Gamma_{i,\mu}(x)$ is only defined on integral x, we can extend the definition to where x can be fractional. This is naturally defined by making $\Gamma_{i,\mu}$ a piecewise linear function. Consider any fractional $x = x' + \lambda$ where $\lambda \in (0, 1)$ and x' is fractional. We assume that $\Gamma_{i,\mu}(x) = (1 - \lambda)\Gamma_{i,\mu}(x') + \lambda\Gamma_{i,\mu}(x' + 1)$. Note that this preserves convexity and concavity of $\Gamma_{i,\mu}$. In this case, [10] has shown that an online $O(c)$-competitive algorithm that fractionally assigns processors to jobs can be converted to an $O(c)$-competitive online algorithm that integrally assigns processors to the jobs without knowing the functions $\Gamma_{i,\mu}$. In short, one can simulate fractional processor assignments by assigning jobs to an additional processor for a small amount of time, leading to equivalent speedup. Thus, we assume WLOG that our algorithm can assign processors fractionally to the jobs. Furthermore, by scaling $\Gamma_{i,\mu}$, we may assume that $m = 1$ throughout the analysis.

We will also extend this problem definition to allow phases of jobs J_i^μ where $\Gamma_{i,\mu}(x) = 1$ *for all* $x \in [0, \infty)$. We note that this implies that phase μ of job i, J_i^μ is processed even though no processor is assigned to it. Although this is not realistic in practice, this only makes our problem harder, yet, in fact, it will help to simplify our analysis. We will call such phases **sequential**. Note that in this case, these phases are not explicitly identified to the algorithm (since it is convexity-sensitive and does not know the functions $\Gamma_{i,\mu}$), so the algorithm may waste processing power on them.

The following simple proposition about concave and convex functions will be useful throughout the analysis.

Proposition 1. – *For any positive value x, any positive $\alpha < 1$, and any concave function f where $f(0) = 0$, we have that $\alpha f(x) \leq f(\alpha x)$.*
- *Also, for any values a and b where $b \geq a > 0$, we have that $\frac{f(b)}{f(a)} \leq \frac{b}{a}$.*
- *Likewise, for any positive value x, any positive value $\alpha < 1$, and any convex function g where $g(0) = 0$, we have that $\alpha g(x) \geq g(\alpha x)$.*
- *Also, we have that for $b \geq a > 0$ $\frac{g(b)}{g(a)} \geq \frac{b}{a}$.*

2.1 Amortized Local Competitiveness

We let $F^A(I)$ and $F^O(I)$ refer to the objective values of the algorithm and OPT on input I. A scheduling algorithm is said to be **d-competitive** if

$$\max_I \frac{F^A(I)}{F^O(I)} \leq d.$$

To prove the competitiveness of an online algorithm, we use an *amortized local competitiveness argument*. See [14] for a tutorial on this technique. To incorporate such an argument it suffices to show for an algorithm A that a potential function $\Phi(t)$ with the following properties exists, where $\Phi(t)$ is continuous at all times except possibly when jobs arrive or are completed:

Boundary Condition: Before any job is released $\Phi(0) = 0$, and after all jobs are finished $\Phi(\infty) \geq 0$.

Completion Condition: Summing over all job completions by the optimal solution and the algorithm, Φ does not increase by more than $\alpha_1 F^O$ for some $\alpha_1 \geq 0$.

Arrival Condition: Summing over the arrival of all jobs, Φ does not increase by more than $\alpha_2 F^O$ for some $\alpha_2 \geq 0$.

Running Condition: At any time t when no job arrives nor is completed,

$$\frac{\partial F^A(t)}{\partial t} + \frac{\partial \Phi(t)}{\partial t} \leq \alpha_3 \frac{\partial F^O(t)}{\partial t}. \tag{1}$$

By integrating Eq. 1 over time and applying the boundary, arrival, and completion conditions we get

$$F^A - \Phi(0) + \Phi(\infty) \leq (\alpha_1 + \alpha_2 + \alpha_3)F^O \tag{2}$$
$$F^A \leq (\alpha_1 + \alpha_2 + \alpha_3)F^O - \Phi(\infty).$$

Equation 2 implies that algorithm A is $(\alpha_1 + \alpha_2 + \alpha_3)$-competitive.

3   A Convexity-Sensitive Scheduling Algorithm

In this section, we present a *convexity-sensitive* scheduling algorithm that is $O(1/\lambda\varepsilon^4)$-competitive with $(1 + \lambda\varepsilon)$-speed augmentation for every $0 < \varepsilon \le 1/2$ and $\lambda > 7/3$. Note that ε can be set independent of λ. For the remaining portion of the algorithm definition and analysis, fix a pair of such λ and ε.

We introduce some definitions and then present our scheduling algorithm.

Definition 1. *We say that phase μ of job i, J_i^μ is **sequential** if $\forall x \ge 0$, $\Gamma_{i,\mu}(x) = 1$. A **linear** phase of job j is a phase J_j^μ with speed-up function $\Gamma_{j,\mu}(x) = x$. We define both of these phases to be concave.*

Definition 2. *Fix a unit time step at time t. For a given scheduling algorithm, we let $A(t)$ be the set of unsatisfied jobs at time t. Also let $A'(t)$ be the $\varepsilon A(t)$ latest arriving jobs. Let $(1 - \gamma_t)$ be the fraction of alive jobs that are currently in a convex phase at time t and a γ_t fraction are in a concave phase at time t.*

Algorithm 1. The Convexity-sensitive Scheduling Algorithm.

1: On each time step t do the following:
2: Give each of the $(1 - \gamma_t)|A'(t)|$ jobs in a convex phase *all of the processing units* for $1/|A'(t)|$ fraction of the unit time slot.
3: Give each of the jobs in a concave phase $1/(\gamma_t|A'(t)|)$ of the processing units for a γ_t fraction of the unit time slot.

---

4   Proof of Theorem 1

We first define our potential function in Subsect. 4.1. Then, in Subsect. 4.2 we argue the *amortized local competitiveness* conditions for our potential function, proving that the convexity-sensitive scheduling algorithm is $O(1)$-competitive.

4.1   The Potential Function Φ

Definition 3. *Let $p_{i,\mu}^A(t)$, and $p_{i,\mu}^O(t)$ be the remaining processing time for J_i's μ-th phase in the algorithm's schedule, and OPT's schedule at time t respectively.*

If job J_i is processed using x_i processing units by the algorithm when it is in phase μ in the algorithm's schedule at time t then $p_{i,\mu}^A(t)$ decreases at a rate of $(1 + \lambda\varepsilon)\Gamma_{i,\mu}(x_i)$. Similarly, if job J_i is processed using x_i processing units by the optimal solution in phase μ at time t then $p_{i,\mu}^O(t)$ decreases at a rate of $\Gamma_{i,\mu}(x_i)$

Definition 4. *We define the **lag** of the algorithm on job J_i's μ-th phase, J_i^μ, compared to OPT as following:*

$$z_{i,\mu}(t) = \max\{p_{i,\mu}^A(t) - p_{i,\mu}^O(t), 0\}.$$

Note that $z_{i,\mu}$ is never negative.

Definition 5. *We define the **rank of job** J_i, $rank_i(t)$, as the number of jobs in $A(t)$ that arrived before job J_i in the system.*

Definition 6. *We define $\beta_{i,\mu}(x)$ to be equal to*

$$\beta_{i,\mu}(x) = \begin{cases} \Gamma_{i,\mu}(1/\varepsilon x) & \text{if } \Gamma_{i,\mu} \text{ is concave;} \\ \Gamma_{i,\mu}(1)/\varepsilon x & \text{if } \Gamma_{i,\mu} \text{ is convex.} \end{cases}$$

Finally, we are ready to define our potential function.

Definition 7. *Let $c = 20/\varepsilon^2(9\lambda - 21)$. We define the potential function $\Phi(t)$*

$$\Phi(t) = c \sum_{i \in A(t)} \sum_{\mu \in \chi_i} \frac{z_{i,\mu}(t)}{\beta_{i,\mu}(rank_i(t))}.$$

4.2 Amortized Local Competitiveness of Φ

In this subsection, we show that our potential function Φ satisfies the four conditions laid out in the Sect. 2 and thus the convexity-sensitive scheduling algorithm is constant competitive.

The following lemma shows that the algorithm has the first three conditions. Its proof is rather straightforward and is omitted due to space constraints.

Lemma 1. *The potential function, $\Phi(t)$, satisfies the* boundary, arrival *and* completion *conditions. In particular, the potential does not increase at any of these events.*

To prove Theorem 1, we need to show the **running condition** for Φ. To do this, we show that we can focus on certain classes of problem instances. The following is a simple extension of a lemma proven in [9].

Lemma 2. *Let S be a* convexity-sensitive scheduler *with s-speed augmentation. Let I be an instance of jobs with phases that have concave or convex speed up functions. There exists an instance I' that includes the same set of convex phases for jobs in I, and for every concave phase J_i^μ in I' it is the case that either J_i^μ is sequential ($\Gamma_{i,\mu} = 1$), or linear ($\Gamma_{i,\mu}(x) = x$). Furthermore, such an I' exists where $F^S(I) = F^S(I')$ and $F^O(I') \le F^O(I)$ where F^O is the objective of the optimal solution and F^S is the objective of the convexity-sensitive scheduler.*

The proof of the above lemma is implied immediately by the proof of Lemma 3.1 in [9]. Specifically, the proof proceeds constructively, by substituting each con-cave phase that is not sequential or linear with one that is; in each case, the objective of the optimal solution decreases, while that of the convexity-sensitive scheduler stays the same. Our construction does not change the convex phases.

This lemma implies that if an instance has concave phases that are not sequential or linear, the performance of our algorithm can only improve rela-tive to the optimal solution. Thus we assume throughout the proof that concave phases are either sequential or linear.

Definition 8. *Let $U(t)$ be the set of unsatisfied jobs in the optimal solution, OPT, at time t.*

First we show how much Φ can increase at time t due to the optimal solution processing jobs. Then, later, we bound the decrease due to the algorithm processing separately.

Recall that for total flow time, the increase in the objective at any point in time is the number of unsatisfied jobs. Thus, for an instantaneous time t we have

$$\frac{\partial F^A(t)}{\partial t} = |A(t)|, \quad \frac{\partial F^O(t)}{\partial t} = |U(t)|.$$

Lemma 3. *The total increase in $\Phi(t)$ at any time t due to the optimal solution processing jobs is at most $c|U(t)| + c\varepsilon|A(t)|$.*

Proof. Our goal is to show that the optimal solution cannot increase the potential function too much. To show this, consider the number of processing units the optimal solution assigns to the jobs.

Definition 9. *Let $m_i^O(t)$ be the number of processing units OPT assigns to job J_i at time t.*

Let $\mu_i^O(t)$ be the phase of job J_i in OPT's schedule at time t. Job J_i is processed by OPT at the rate of $\Gamma_{i,\mu_i^O(t)}(m_i^O(t))$ (the remaining processing time $p_{i,\mu_i^O(t)}^O(t)$ of phase $\mu_i^O(t)$ for job i would *decrease* at this rate). Then $z_{i,\mu_i^O(t)}(t)$ could *increase* by this amount in the worst case. Hence,

$$\frac{\partial \Phi(t)}{\partial t} \le c \sum_{i \in U(t)} \frac{\Gamma_{i,\mu_i^O(t)}(m_i^O(t))}{\beta_{i,\mu_i^O(t)}(\mathrm{rank}_i(t))}.$$

Let $U_v(t)$ be the alive jobs in OPT at time t that are in a *convex* phase at time t in OPT's schedule and $U_c(t)$ be ones that are in a *concave* phase.

$$
\begin{aligned}
\frac{\partial \Phi(t)}{\partial t} &\le c \sum_{i \in U(t)} \frac{\Gamma_{i,\mu_i^O(t)}(m_i^O(t))}{\beta_{i,\mu_i^O(t)}(\mathrm{rank}_i(t))} \\
&\le c \left(\sum_{i \in U_c(t)} \frac{\Gamma_{i,\mu_i^O(t)}(m_i^O(t))}{\beta_{i,\mu_i^O(t)}(\mathrm{rank}_i(t))} + \sum_{i \in U_v(t)} \frac{\Gamma_{i,\mu_i^O(t)}(m_i^O(t))}{\beta_{i,\mu_i^O(t)}(\mathrm{rank}_i(t))} \right) \\
&\le c \left(\sum_{i \in U_c(t)} \frac{\Gamma_{i,\mu_i^O(t)}(m_i^O(t))}{\Gamma_{i,\mu_i^O(t)}(1/\varepsilon\mathrm{rank}_i(t))} + \varepsilon \sum_{i \in U_v(t)} \frac{\mathrm{rank}_i(t)\Gamma_{i,\mu_i^O(t)}(m_i^O(t))}{\Gamma_{i,\mu_i^O(t)}(1)} \right).
\end{aligned}
$$

We now bound each of these terms separately.

First Summation Term. If $m_i^O(t) \le 1/(\varepsilon\mathrm{rank}_i(t))$, then the corresponding term in the first summation is at most 1. Note that this is true even if the job is in a sequential phase and we can assume OPT does not assign any processors to sequential jobs since it does not increase the rate they are processed.

On the other hand, if $m_i^O(t) > 1/(\varepsilon \mathbf{rank}_i(t))$, then by Proposition 1, we have

$$\frac{\Gamma_{i,\mu_i^O(t)}(m_i^O(t))}{\Gamma_{i,\mu_i^O(t)}(1/\varepsilon \mathbf{rank}_i(t))} \leq \varepsilon \mathbf{rank}_i(t) m_i^O(t).$$

due to the concavity of any job J_i's phase in the first sum. Therefore,

$$\sum_{i \in U_c(t)} \frac{\Gamma_{i,\mu_i^O(t)}(m_i^O(t))}{\Gamma_{i,\mu_i^O(t)}(1/\varepsilon \mathbf{rank}_i(t))} \leq |U_c(t)| + \sum_{i \in U_c(t)} \varepsilon \mathbf{rank}_i(t) m_i^O(t).$$

Second Summation Term. As for the second term we know that $m_i^O(t) \leq 1$. Hence, due to the convexity of the jobs phases, Proposition 1 implies that

$$\frac{\Gamma_{i,\mu_i^O(t)}(m_i^O(t))}{\Gamma_{i,\mu_i^O(t)}(1)} \leq m_i^O(t).$$

Thus, we get that

$$\sum_{i \in U_v(t)} \frac{\mathbf{rank}_i(t)\Gamma_{i,\mu_i^O(t)}(m_i^O(t))}{\Gamma_{i,\mu_i^O(t)}(1)} \leq \sum_{i \in U_v(t)} \mathbf{rank}_i(t) m_i^O(t).$$

Substituting the above simplifications, we get

$$\frac{\partial \Phi(t)}{\partial t} \leq c\varepsilon \mathbf{rank}_i(t) \left(\sum_{i \in U_c(t)} m_i^O(t) + \sum_{i \in U_v(t)} m_i^O(t) \right) + c|U_c(t)|$$

$$\leq c|U_c(t)| + c\varepsilon|A(t)| \sum_{i \in U(t)} m_i^O(t).$$

The last line follows because the rank of each job is at most $A(t)$ at time t, by definition of the ranks of the jobs and because there are only $A(t)$ unsatisfied jobs in the algorithm's schedule at time t.

Finally, we know that $\sum_{i \in U(t)} m_i^O(t) \leq 1$, because the amount of processing units divided between jobs at any time can not exceed $m = 1$. Hence,

$$\frac{\partial \Phi(t)}{\partial t} \leq c|U_c(t)| + c\varepsilon|A(t)| \leq c|U(t)| + c\varepsilon|A(t)|.$$

$\square$

We divide the rest of the analysis into two cases depending on the relationship between $U(t)$ and $A(t)$. First, we consider the easier case where $|U(t)| \geq \varepsilon^2|A(t)|/10$.

Lemma 4. *If $|U(t)| \geq \varepsilon^2|A(t)|/10$, then $\partial F^A(t)/\partial t + \partial \Phi(t)/\partial t$ is at most $O(1/\lambda \varepsilon^4)\left(\partial F^O(t)/\partial t\right)$.*

Proof. By Lemma 3 we know that the most OPT can increase $\Phi(t)$ is $c|U(t)| + c\varepsilon|A(t)|$. Hence,

$$c|U(t)| + c\varepsilon|A(t)| \leq c(1 + 10/\varepsilon)|U(t)| \qquad \text{since } \varepsilon|A(t)| \leq 10|U(t)|/\varepsilon.$$

Therefore we have,

$$\frac{\partial F^A(t)}{\partial t} + \frac{\partial \Phi(t)}{\partial t} \leq A(t) + c(1 + 10/\varepsilon)|U(t)|$$

$$\leq c(1 + 10/\varepsilon + 10/\varepsilon^2)|U(t)| \qquad \text{since } |A(t)| \leq 10|U(t)|/\varepsilon^2,$$

$$\leq O(1/\lambda\varepsilon^4)\left(\frac{\partial F^O(t)}{\partial t}\right) \qquad \text{since } c = \frac{20}{\varepsilon^2(9\lambda - 21)}.$$

Now we consider the more challenging case where $|U(t)| < \varepsilon^2|A(t)|/10$.

Lemma 5. *If $|U(t)| < v^2|A(t)|/10$, then $\partial F^A(t)/\partial t + \partial \Phi(t)/\partial t$ is at most $O(1/\lambda\varepsilon^4)\left(\partial F^O(t)/\partial t\right)$.*

Proof. In the proof of Lemma 4 we focused on how OPT can increase Φ (by assuming that the algorithm didn't decrease the $p^A_{i,\mu^O_i(t)}(t)$ variables at all). Let $\mu^A_i(t)$ be the phase job J_i is in at time t in the algorithm's schedule. In this proof, we focus on how the algorithm can decrease $z_{i,\mu^A_i(t)}(t)$ and thus $\Phi(t)$ as well as OPT increasing $\Phi(t)$. Let $C^O(t)$ be the change in $\Phi(t)$ due to the optimal solution processing jobs at time t and let $C^A(t)$ denote the change in $\Phi(t)$ due to the algorithm processing of jobs at time t. Lemma 3 says that $C^O(t) \leq c|U(t)| + c\varepsilon|A(t)|$.

Now, we bound $C^A(t)$ at a time t where $|U(t)| < \varepsilon^2|A(t)|/10$. Recall that $z_{i,\mu^A_i(t)}(t) = \max\{p^A_{i,\mu^A_i(t)}(t) - p^O_{i,\mu^A_i(t)}(t), 0\}$. Therefore, $z_{i,\mu^A_i(t)}$ can only decrease due to the algorithm's processing. Further, $z_{i,\mu^A_i(t)}(t)$ will decrease at the rate the algorithm process job J_i at time t if the optimal solution has completed J_i by time t. That is, for jobs not in $U(t)$. Since OPT only has $U(t) < \varepsilon^2|A(t)|/10$ unfinished jobs, the algorithm's processing on at least a $(1 - \varepsilon/10)$ fraction of the jobs in $A'(t)$ causes $z_{i,\mu^A_i(t)}$ to decrease at the rate they are processed.

Let $A_c(t)$ be the set of jobs in $A'(t)$ that are in a concave phase at time t in the algorithm's schedule. Let $A_v(t)$ be the set of jobs in $A'(t)$ that are in a convex phase at time t in the algorithm's schedule. $A'(t) = A_c(t) \cup A_v(t)$. We assume that at time t a $(1 - \gamma_t)$ fraction of jobs in $A'(t)$ are in $A_v(t)$ and a γ_t fraction are in $A_c(t)$

By Lemma 2, we may assume that any concave phase for a job is either *sequential* or *linear*. Therefore, to bound $C^A(t)$ we can assume that all the jobs in $A_c(t)$ are either sequential or linear. Let $S_c(t)$ be the jobs in $A_c(t)$ that are in a sequential phase at time t and the others are in $L_c(t)$. $A_c(t) = S_c(t) \cup L_c(t)$.

Here are the progress rates of the algorithm on $A_v(t)$, $S_c(t)$, and $L_c(t)$.

- The algorithm processes each convex job in $A_v(t)$ at a rate of $\Gamma_{i,\mu_i^A(t)}(1)/\varepsilon|A(t)|$.
- For each linear job in $L_c(t)$, the rate of progress is $\gamma_t\Gamma_{i,\mu_i^A(t)}(1/\gamma_t\varepsilon|A(t)|) = 1/\varepsilon|A(t)|$.
- For each sequential job in $S_c(t)$, the rate of progress is always 1 no matter how many processing units are assigned to the job (even if there are 0 units assigned).

The algorithm is $(1+\lambda\varepsilon)$-speed augmented; therefore we multiply its change in the potential function by $(1+\lambda\varepsilon)$. Combining all the above, we get

$$C(t)^A \leq -c(1+\lambda\varepsilon)\left(\sum_{i\in A_v(t)\setminus U(t)} \frac{\Gamma_{i,\mu_i^A(t)}(1)/\varepsilon|A(t)|}{\beta_{i,\mu_i^A(t)}(\mathbf{rank}_i(t))} + \sum_{i\in S_c(t)\setminus U(t)} \frac{1}{\beta_{i,\mu_i^A(t)}(\mathbf{rank}_i(t))} \right.$$
$$\left. + \sum_{i\in L_c(t)\setminus U(t)} \frac{1/\varepsilon|A(t)|}{\beta_{i,\mu_i^A(t)}(\mathbf{rank}_i(t))} \right). \tag{3}$$

Note that since the algorithm works on the latest $A'(t)$ arriving jobs, the rank of each job in $A'(t)$ is bounded between

$$(|A(t)| - |A'(t)|) \leq \mathbf{rank}_i(t) \leq |A(t)|$$
$$(1-\varepsilon)|A(t)| \leq \mathbf{rank}_i(t) \leq |A(t)|.$$

By starting from Inequality 3, replacing the definition of $\beta_{i,\mu_i^A(t)}$, and using the above bounds on the rank of each job in $A'(t)$, we can show the following proposition. The proof is omitted.

Proposition 2. *Let $C^A(t)$ be the change in $\Phi(t)$ due to the algorithm processing of jobs at time t. For $\varepsilon < 1/2$ and $|U(t)| < \varepsilon^2|A(t)|/10$ we have*

$$C^A(t) \leq -c\varepsilon|A(t)|\left(1 + (9\lambda - 21)\frac{\varepsilon}{20}\right).$$

Using Proposition 2, we can upper bound $\partial\Phi(t)/\partial t$.

$$\frac{\partial\Phi(t)}{\partial t} = C^O(t) + C^A(t)$$

$$\leq (c|U(t)| + c\varepsilon|A(t)|) - \left(c\varepsilon|A(t)| + c\varepsilon^2\left(\frac{9\lambda - 21}{20}\right)|A(t)|\right)$$

$$\leq c|U(t)| - c\varepsilon^2\left(\frac{9\lambda - 21}{20}\right)|A(t)|.$$

With this we get the **running** condition (Eq. 1) which finishes the proof of Lemma 5:

$$\frac{\partial F^A(t)}{\partial t} + \frac{\partial\Phi(t)}{\partial t} \leq c|U(t)| + |A(t)|\left(1 - c\varepsilon^2\left(\frac{9\lambda - 21}{20}\right)\right)$$

$$\leq c|U(t)| = c\left(\frac{\partial F^O(t)}{\partial t}\right) \quad \text{since } c = \frac{20}{\varepsilon^2(9\lambda - 21)}, \ \lambda > 7/3.$$

□

Lemmas 4 and 5 together imply the running condition and complete the proof of Theorem 1.

Acknowledgements. We would like to thank Michael Bender for helpful discussions, and Bertrand Simon for informing us of reference [2]

References

1. Bansal, N., Krishnaswamy, R., Nagarajan, V.: Better scalable algorithms for broadcast scheduling. In: Abramsky, S., Gavoille, C., Kirchner, C., Meyer auf der Heide, F., Spirakis, P.G. (eds.) ICALP 2010. LNCS, vol. 6198, pp. 324–335. Springer, Heidelberg (2010)
2. Beaumont, O., Guermouche, A.: Task scheduling for parallel multifrontal methods. In: Kermarrec, A.-M., Bougé, L., Priol, T. (eds.) Euro-Par 2007. LNCS, vol. 4641, pp. 758–766. Springer, Heidelberg (2007)
3. Blazewicz, J., Kovalyov, M.Y., Machowiak, M., Trystram, D., Weglarz, J.: Preemptable malleable task scheduling problem. IEEE Trans. Comput. **55**(4), 486–490 (2006)
4. Blazewicz, J., Machowiak, M., Weglarz, J., Kovalyov, M.Y., Trystram, D.: Scheduling malleable tasks on parallel processors to minimize the makespan. Ann. Oper. Res. **129**(1–4), 65–80 (2004)
5. Chadha, J.S., Garg, N., Kumar, A., Muralidhara, V.N.: A competitive algorithm for minimizing weighted flow time on unrelated machines with speed augmentation. In: Proceedings of the 41st Symposium on Theory of Computation (STOC) (2009)
6. Chan, S.H., Lam, T.W., Lee, L.K., Zhu, J.: Nonclairvoyant sleep management and flow-time scheduling on multiple processors. In: Proceedings of the 25th Symposium on Parallelism in Algorithms and Architectures (SPAA), pp. 261–270 (2013)
7. Edmonds, J.: Scheduling in the dark. Theoret. Comput. Sci. **235**(1), 109–141 (2000). Preliminary version in STOC 1999
8. Edmonds, J., Im, S., Moseley, B.: Online scalable scheduling for the ℓ_k-norms of flow time without conservation of work. In: Proceedings of the Twenty-Second Annual ACM-SIAM Symposium on Discrete Algorithms (SODA) (2011)
9. Edmonds, J., Pruhs, K.: Scalably scheduling processes with arbitrary speedup curves. ACM Trans. Algorithms **8**(3), 28:1–28:10 (2012)
10. Fox, K., Im, S., Moseley, B.: Energy efficient scheduling of parallelizable jobs. In: Proceedings of the 24th ACM-SIAM Symposium on Discrete Algorithms (SODA), pp. 948–957 (2013)
11. Fox, K., Moseley, B.: Online scheduling on identical machines using SRPT. In: Proceedings of the 22nd ACM Symposium on Discrete Algorithms (SODA) (2011)
12. Gupta, A., Im, S., Krishnaswamy, R., Moseley, B., Pruhs, K.: Scheduling jobs with varying parallelizability to reduce variance. In: Proceedings of the Twenty-Second Syposium on Parallel Algorithms and Architectures (SPAA), pp. 11–20 (2010)
13. Im, S., Moseley, B.: Online scalable algorithm for minimizing ℓ_k-norms of weighted flow time on unrelated machines. In: Proceedings of the Twenty-Second Annual ACM Symposium on Discrete Algorithms (SODA), pp. 95–108 (2011)
14. Im, S., Moseley, B., Pruhs, K.: A tutorial on amortized local competitiveness in online scheduling. SIGACT News **42**, 83–97 (2011)

15. Im, S., Moseley, B., Pruhs, K., Torng, E.: Competitively scheduling tasks with intermediate parallelizability. In: Proceedings of the Twenty-Sixth ACM Symposium on Parallelism in Algorithms and Architectures (SPAA), pp. 22–29 (2014)
16. Kalyanasundaram, B., Pruhs, K.: Speed is as powerful as clairvoyance. J. ACM 47(4), 617–643 (2000)
17. Leonardi, S., Raz, D.: Approximating total flow time on parallel machines. J. Comput. Syst. Sci. 73(6), 875–891 (2007)
18. Ludwig, W., Tiwari, P.: Scheduling malleable and nonmalleable parallel tasks. In: Proceedings of the 5th ACM-SIAM Symposium on Discrete Algorithms (SODA), pp. 167–176 (1994)
19. Prasanna, G.N.S., Musicus, B.R.: Generalized multiprocessor scheduling and applications to matrix computations. IEEE Trans. Parallel Distrib. Syst. 7(6), 650–664 (1996)
20. Pruhs, K., Sgall, J., Torng, E.: Handbook of Scheduling: Algorithms, Models, and Performance Analysis. CRC Press, Boca Raton (2004). Online Scheduling

Scheduling with State-Dependent Machine Speed

Veerle Timmermans[✉] and Tjark Vredeveld

Maastricht University, Maastricht, The Netherlands
{v.timmermans,t.vredeveld}@maastrichtuniversity.nl

Abstract. We study a preemptive single machine scheduling problem where the machine speed is externally given and depends on the number unfinished jobs. The objective is to minimize the sum of weighted completion times. We develop a greedy algorithm that solves the problem to optimality when we work with either unit weights or unit processing times. If both weights and processing times are arbitrary, we show the problem is NP-hard by making a reduction from 3-partition.

1 Introduction

In queueing theory, many studies have been made about queues with state-dependent service speeds, see e.g. Bekker and Boxma [4] or Bekker, Borst, Boxma and Kella [13] and the references therein. This model is among others motivated by Bertrand and Van Ooijen [5] through human servers who may be slow when there is much work to do, due to stress, or when there is little work to do due to laziness. For state-dependent server speeds in packet-switched communication systems, we refer to [6,7,11,14].

Although these models have been extensively studied in queueing theory, not much is known about algorithms that solve these models to optimality nor the computational complexity of this type of problem. During the 2013 Scheduling workshop in Dagstuhl, Urtzi Ayesta [3] posed it as an open question how optimal policies look like and what the computational complexity is. In this paper, we settle this open problem for one variant of state-dependent machine speeds, namely when the speed of the machine varies with the number of jobs in the system. This number of jobs is a good measure for the total workload in the system, when the service requirements of the jobs are i.i.d. Moreover, in this setting the speed of the server only changes at discrete times, see e.g. [4].

1.1 Previous Work

Models with workload dependent server speeds originate from queueing theory. Bertrand and Van Ooijen [5] assume in their paper that the workload level affects the effective processing times in a job shop. This assumption is based on the results of empirical research on the relationship between workload and shop performance. Bekker, Borst, Boxma and Kella [13] consider two types of queues

© Springer International Publishing Switzerland 2015
L. Sanità and M. Skutella (Eds.): WAOA 2015, LNCS 9499, pp. 196–208, 2015.
DOI: 10.1007/978-3-319-28684-6_17

with workload dependent arrival rate and service speed. Bekker and Boxma [4] consider a queueing system where feedback information about the level of congestion is given right after arrival instants. When the amount of work right after arrival is at most some threshold, then the server works at a low speed until the next arrival instant.

Related work in deterministic scheduling with varying machine speeds includes the following. Research into speed scaling algorithms started with the work of Yao, Demers and Shenker [8], where each job is to be executed between its arrival time and deadline by a single processor with variable speed. The difference with the previous mentioned models as well as ours is that the scheduler in this setting also needs to decide on the speed of the machine at any time t. A review paper on speed scaling algorithms is written by Albers [1].

Megow and Verschae [12] also study scheduling problems on a machine of varying speed, but they assume a speed function that depends on the time which is known a priori. They developed a PTAS for minimizing the total weighted completion time.

The machine speed model we consider, was previously investigated by Gawiejnowicz [10] and Alidaee and Ahmadian [2]. Though both papers discussed the non-preemptive case instead of the preemptive case we are looking at. Here Gawiejnowicz considered the makespan objective whereas we study the goal to minimize the sum of (weighted) completion. Alidaee and Ahmadian studied the sum of completion times where jobs have equal weights.

1.2 Problem Definition

In the model under consideration, n jobs need to be scheduled on a single machine. A job j is associated with a strictly positive processing requirement denoted by p_j and depending on the variant that we consider also with a weight w_j. All jobs as well as the machine are available from the beginning. The machine is allowed to preempt a job, i.e., the processing of a job may be interrupted and resumed later on the machine. By allowing infinitesimally small processing on a job before preempting it, the preemption model can be viewed as one in which during each time interval the processing capacity of the machine is divided over one or more jobs. The goal is to minimize the total (weighted) completion time, $\sum w_j C_j$, where C_j denotes the completion time of job j. In our scheduling model, the speed at which the machine processes its jobs varies with the number of jobs in the system. For notational convenience, we represent the speed at which the machine is processing as a function of the number of *completed* jobs. Hereto, we are given a speed function s_i $(i = 1, \ldots, n)$, where s_i denotes the speed of the machine between the $(i-1)$th and ith completion. From now on we refer to this problem as JDMS.

When the machine speed does not vary, i.e. $s_i = 1$ for all i, the problem is a standard problem where the total weighted completion time should be minimized on a single machine. This problem is solved by Smith's rule [15].

1.3 Our Results

When we know the order in which the jobs complete, we can formulate the problem as an LP with two types of constraints. The first type of constraints ensures that the amount of processing done during an interval is not more than the capacity of the machine in that interval. The second type of constraints ensures that the jobs complete in the order that is assumed. Any solution to our scheduling problem can be represented as a solution to the LP and vice versa. Using this fact, we can prove that there exists an optimal solution such that at every completion time C_j, any job k is already completed by time C_j or it has not received any processing yet by this time.

Intuitively this lemma means that there exists an optimal solution where the jobs are partitioned in groups and all jobs in a group start and finish at the same time. Given the order of job completions, we still need to decide on how to partition the jobs into groups. Although this follows from the optimal LP solution, we also develop a combinatorial greedy algorithm. For the case of the total completion time objective, it is easy to show that jobs will complete in shortest processing time (SPT) order, whereas the case of unit processing times the optimal order will be sorting according to non-increasing weight. When both weight and processing times can be arbitrary, the problem is strongly NP-hard and we make a reduction from 3-partition.

2 Structural Property

In this section we show that an optimal schedule has a certain block structure in which the jobs are processed in groups and that the jobs in one group start and finish at the same time. Note that we view the preemption model as one in which the processing capacity of the machine is divided over one or more jobs. Therefore, we may assume that a set of jobs can complete at exactly the same time. In case that the order in which the jobs need to complete is given, we can formulate the problem of minimizing the total weighted completion time as a linear program. Assuming w.l.o.g. that the order of completion is given by the index of jobs, i.e., $C_1 \leq \cdots \leq C_n$, the variables in the LP denote the time between the ith and the $i + 1$st completion. That is, we use variable Δ_i, where:

$$\Delta_i = \begin{cases} C_1 & \text{if } i = 1, \\ C_i - C_{i-1} & \text{if } 1 < i \leq n. \end{cases}$$

Hence, the completion time of job j can be written as $C_j = \sum_{i=1}^{j} \Delta_i$. At the interval $[0, C_1]$, the machine is operating at speed s_1 and during the intervals $[C_{i-1}, C_i]$, the machine is operating at speed s_i. The sum of weighted completion times can be rewritten as:

$$\sum_{j=1}^{n} w_j C_j = \sum_{j=1}^{n} w_j \sum_{i=1}^{j} \Delta_i = \sum_{i=1}^{n} \left(\sum_{j=i}^{n} w_j \cdot \Delta_i \right).$$

To make sure the requested order on the completion times is enforced, we have the constraint $\Delta_i \geq 0$ for all i. Lastly we want to make sure that by time C_i at least jobs $1, \ldots, i$ have been fully processed. Thus the total amount of processing up to time C_i needs to be at least the processing requirements of the first i jobs:

$$\sum_{k=1}^{i} \Delta_k \cdot s_k \geq \sum_{k=1}^{i} p_k, \qquad 0 \leq i \leq n.$$

The LP is as follows:

$$\text{minimize} \sum_{i=1}^{n} \left(\sum_{j=i}^{n} w_j \cdot \Delta_i \right)$$

$$\text{subject to} \sum_{k=1}^{i} \Delta_k \cdot s_k \geq \sum_{j=1}^{i} p_j, \qquad 1 \leq i \leq n$$

$$\Delta_i \geq 0, \qquad\qquad 1 \leq i \leq n.$$

Note that a feasible solution for this LP does not correspond to a unique schedule, but with a non-empty set of schedules for which the completion times are set. Any feasible schedule leads to a unique solution of the LP. For any given order on the completion times, we can find an optimal schedule in polynomial time using this LP.

We use this LP to show that an optimal solution to the general JDMS has a block structure. We first prove that there exists an optimal solution, such that at all completion times, each job that has started is finished.

Lemma 1. *There exists an optimal schedule such that at every completion time C_i, for every job j one of the following holds:*

1. *Job j is already completed at time C_i.*
2. *At time C_i job j has not received any processing yet.*

Proof. Given an order of completion times, we make a corresponding LP and reformulate this lemma in terms of this linear program. For this linear program we prove that for each job k one of the following holds:

- $\Delta_{k+1} = 0$, and thus $C_k = C_{k+1}$.
- If $C_{k+1} > C_k$, then it has to hold that $\sum_{j=1}^{k} s_j \Delta_j = \sum_{j=1}^{k} p_j$. If that is not the case, we use the fact that a solution must satisfy the requested order on the completion times in combination with the assumption that a machine is always working at full speed. Then there exists a job $l \geq k+1$ that is not yet completed, but already received some processing.

Thus we want to prove that for $k = 1, \ldots, n-1$ either:

$$\Delta_{k+1} = 0 \quad (1) \qquad \text{or} \qquad \sum_{j=1}^{k} s_j \Delta_j = \sum_{j=1}^{k} p_j. \quad (2)$$

Suppose we have an optimal solution σ such that there exists a k with:

$$\Delta_{k+1} > 0 \quad \text{and} \quad \sum_{j=1}^{k} s_j \Delta_j > \sum_{j=1}^{k} p_j.$$

We define ℓ as:

$$\ell = \max\{j \leq k | \Delta_j > 0\}.$$

Note that ℓ exists as prosessing times are strictly positive. We define two new feasible solutions, for some $\epsilon > 0$:

1. We define σ' as the solution where $\Delta_\ell^{\sigma'} = \Delta_\ell^{\sigma} - \frac{\epsilon}{s_\ell}$ and $\Delta_{k+1}^{\sigma'} = \Delta_{k+1}^{\sigma} + \frac{\epsilon}{s_{k+1}}$. The change in the objection value is:

$$\sum_{i=1}^{n} w_i C_i^{\sigma'} - \sum_{i=1}^{n} w_i C_i^{\sigma} = \sum_{1 \leq i \leq j \leq n} w_j \Delta_i^{\sigma'} - \sum_{1 \leq i \leq j \leq n} w_j \Delta_i^{\sigma}$$

$$= \left(\sum_{i=k+1}^{n} w_i \right) \frac{\epsilon}{s_{k+1}} - \left(\sum_{i=\ell}^{n} w_i \right) \frac{\epsilon}{s_\ell}.$$

Note that σ' is still feasible, as we can do ϵ amount of work less in $\Delta_\ell^{\sigma'}$ compared to Δ_ℓ^{σ}, but ϵ amount of work extra in $\Delta_{k+1}^{\sigma'}$ compared to Δ_{k+1}^{σ}.

2. We define σ'' as the solution where $\Delta_\ell^{\sigma''} = \Delta_\ell + \frac{\epsilon}{s_\ell}$ and $\Delta_{k+1}^{\sigma''} = \Delta_{k+1} - \frac{\epsilon}{s_{k+1}}$. The change in the objection value is:

$$\sum_{i=1}^{n} w_i C_i^{\sigma''} - \sum_{i=1}^{n} w_i C_i^{\sigma} = \sum_{1 \leq i \leq j \leq n} w_j \Delta_i^{\sigma''} - \sum_{1 \leq i \leq j \leq n} w_j \Delta_i^{\sigma}$$

$$= \left(\sum_{i=\ell}^{n} w_i \right) \frac{\epsilon}{s_\ell} - \left(\sum_{i=k+1}^{n} w_i \right) \frac{\epsilon}{s_{k+1}}.$$

Note that σ'' is still feasible, as we can do ϵ amount of work extra in $\Delta_\ell^{\sigma''}$ compared to Δ_ℓ^{σ}, but ϵ amount of work less in $\Delta_{k+1}^{\sigma''}$ compared to Δ_{k+1}^{σ}.

As:

$$\sum_{i=1}^{n} w_i C_i^{\sigma'} - \sum_{i=1}^{n} w_i C_i^{\sigma} = - \left(\sum_{i=1}^{n} w_i C_i^{\sigma''} - \sum_{i=1}^{n} w_i C_i^{\sigma} \right),$$

at least one of the two solutions is better than or equal to σ. As σ is optimal, we actually know that both new solutions are also optimal.

Suppose k is the smallest value such that neither (1) or (2) holds. Let:

$$\epsilon = \min\{\Delta_{k+1}, \sum_{j=1}^{k} s_j \Delta_j - \sum_{j=1}^{k} p_j\},$$

then either σ' or σ'' will give an optimal solution where either (1) or (2) holds for job k. We repeat this procedure until this property holds for all $k \in \{0 \ldots n\}$.

Thus given an order on the completion times, we can find an optimal schedule satisfying the requested order such that for all i, j, where $1 \leq i, j \leq n$ it holds that at time C_j either job i is completed or did not receive any processing time yet. As this holds for any order, this will also hold for the order of some optimal solution. Therefore there exists an optimal solution such that or all i, j, where $1 \leq i, j \leq n$, it holds that $y_i(C_j) \in \{0, p_i\}$. $\square$

Lemma 1 implies that we can divide the jobs into groups of consecutive jobs, such that all jobs in a group will start and end at the same time. We use G_i to denote the ith group of jobs and we denote an optimal solution as $[G_1, \ldots, G_k]$ (Fig. 1).

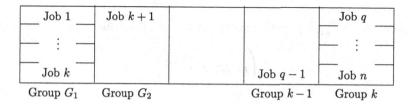

Fig. 1. Schedule in which the jobs are divided into k groups.

3   JDMS is Strongly NP-hard

From the previous section it follows that once we know the order in which the jobs complete in an optimal solution, we can find such an optimal solution in polynomial time. The only question that remains is how to find an order such that there exists an optimal solution that satisfies this order? When there are no restrictions on weights and processing times, it is strongly NP-hard to find such an order.

Theorem 1. *JDMS is strongly NP-hard.*

Proof. We make a reduction from 3-partition [9] to JDMS. We take an instance of 3-partition with $3m$ elements of size a_j, and $B = \frac{1}{m} \sum_{i=1}^{3m} a_j$. The reduction to JDMS is as follows: we define $3m$ jobs, with $p_j = w_j = a_j$. The machine speed is 1 when there are 0 (mod 3) jobs completed, and 0 otherwise. Now we claim that there is a solution for 3-partition if and only if we can find a schedule where the sum of weighted completion times is at most $\frac{1}{2}m(m+1)B^2$.

Suppose we have a yes-instance for 3-partition, then there are sets $S_1, \ldots, S_m$ that contain exactly three elements, and $\sum_{i \in S_j} a_j = B$ for all j. Then for JDMS we make the schedule where we process the jobs in groups $[S_1, \ldots S_m]$. In this schedule the machine runs always at speed 1, as we always process three jobs at the same time. Group S_i has weight B and completion time iB, thus the objective value is in this case $\sum_{i=1}^{m} iB^2$, which equals $\frac{1}{2}m(m+1)B^2$. Thus indeed there exists a schedule with completion time at most $\frac{1}{2}m(m+1)B^2$.

Suppose we have a schedule in JDMS with objective value at most $\frac{1}{2}m(m+1)B^2$. According to Lemma 1 the jobs are processed in groups, and the sizes of these groups have to be a multiples of three. Therefore we have at most m groups. Now let x_i be the length of group i, then the total weighted cost is:

$$\sum_{i=1}^{m} \left(x_i * \sum_{1 \leq j \leq i} x_j \right).$$

This can be rewritten as:

$$\frac{1}{2} \left((\sum_{i=1}^{m} x_i)^2 + \sum_{i=1}^{m} x_i^2 \right).$$

We know that $\sum_{i=1}^{m} x_i = mB$, thus we have objective value:

$$\frac{1}{2} \left(m^2 B^2 + \sum_{i=1}^{m} x_i^2 \right).$$

The sum of squares is minimized when $x_1 = \cdots = x_n = \frac{mB}{m} = B$. Note that this will give us exactly objective value $\frac{1}{2}m(m+1)B^2$, and is therefore the only schedule we could have found. This implies that every group has exactly processing time B. Note that we need to use all m groups, and we know that every group has to contain a multiple of 3 jobs. As we have $3m$ jobs in total, every group has to contain exactly three jobs. Thus these groups form a 3-partition of the jobs a_j.

This implies that the decision variant of JDMS is NP-complete, and therefore JDMS is NP-hard.   □

4   Combinatorial Algorithm for Special Cases

In the previous section we proved that JDMS is strongly NP-hard. As the partitioning problem can be solved within polynomial time, the problem lies in finding a suitable order on the job completions. In some special cases, when working with either unit weights or unit processing requirements, such an order can be found within polynomial time.

4.1   Solving the Sequencing Problem in Special Cases

In the first special case we assume all jobs have an equal weight.

Theorem 2. *Suppose $w_1 = \cdots = w_n$, then there exists an optimal schedule for JDMS in which the jobs are completed in order of non-decreasing processing time.*

Proof. Assume w.l.o.g that $p_1 \leq \cdots \leq p_n$ and we have an optimal schedule σ for which it does not hold that $C_1^\sigma \leq \cdots \leq C_n^\sigma$. Then we look at the smallest i such that $C_{i+1}^\sigma < C_i^\sigma$. We change σ to σ^* by only changing σ at job i and $i+1$. We let job i finish at time $C_i^{\sigma^*} = C_{i+1}^\sigma$ and job $i+1$ finish at time $C_{i+1}^{\sigma^*} = C_i^\sigma$. As $p_i \leq p_{i+1}$ and $C_{i+1}^\sigma < C_i^\sigma$, we know that there needs to be a time t at which in schedule σ the remaining processing time of a job i is the same as the remaining processing time of job $i+1$. We define the schedule σ^* as follows. It processes all jobs $j \neq i, i+1$ the same as in σ as it does with jobs i and $i+1$ up to time t. From time t onwards, σ^* processes job $i+1$ whenever σ processes job i and it processes job i whenever job $i+1$ is processed by σ. All other jobs are processed in σ^* the same way as in σ.

As the time points where jobs finish stay the same, it holds that the speed of the machine is equal in σ and σ^* at every point in time. Therefore per time unit the same amount of work can be done in σ and σ^*. So there is exactly enough space for job $i+1$ to be finished at time C_i^σ. Thus $\sum_{i=1}^n C_i^\sigma = \sum_{i=1}^n C_i^{\sigma^*}$ and σ^* will remain optimal.

Iterating this process implies that there is an optimal schedule σ' such that $C_1^{\sigma'} \leq \cdots \leq C_n^{\sigma'}$. □

Thus for JDMS with equal weights there exists an optimal solution that satisfies the SPT order.

When we work with unit processing requirements instead of weights, we again can find an order on the completion times that guarantees an optimal solution.

Theorem 3. *Suppose $p_1 = \cdots = p_n$, then there exists an optimal schedule for JDMS in which the jobs completed in order of non-increasing weights*

Proof. Suppose we have an optimal schedule σ and it does not hold that $C_1^\sigma \leq \cdots \leq C_n^\sigma$. Then we look at the smallest i such that $C_{i+1}^\sigma < C_i^\sigma$ and $w_i > w_{i+1}$ (if $w_i = w_{i+1}$ then job i and $i+1$ are identical and therefore switching job i with job j will result in an optimal schedule as well). We change σ to σ^* by processing job $i+1$ whenever σ processes job i and job i whenever σ processes job $i+1$. All other jobs are processed as in σ. As $p_i = p_{i+1}$, we know that $C_{i+1}^{\sigma^*} = C_i^\sigma$ and $C_i^{\sigma^*} = C_{i+1}^\sigma$ and $C_j^{\sigma^*} = C_j^\sigma$ for all $j \neq i, i+1$. Thus $w_j C_j^\sigma = w_j C_j^{\sigma^*}$ for all $j \neq i, i+1$. Furthermore, as $w_i > w_{i+1}$, $C_{i+1}^\sigma > C_i^{\sigma^*}$, $C_i^\sigma = C_{i+1}^{\sigma^*}$ and $C_{i+1}^\sigma = C_i^{\sigma^*}$, some simple rewriting learns us that:

$$w_i C_i^\sigma + w_{i+1} C_{i+1}^\sigma = w_{i+1} C_i^{\sigma^*} + w_i C_{i+1}^{\sigma^*}$$
$$= w_{i+1} C_i^{\sigma^*} + w_i C_{i+1}^{\sigma^*} - (w_i C_i^{\sigma^*} + w_{i+1} C_{i+1}^{\sigma^*})$$
$$+ (w_i C_i^{\sigma^*} + w_{i+1} C_{i+1}^{\sigma^*})$$
$$= (w_i - w_{i+1})(C_{i+1}^{\sigma^*} - C_i^{\sigma^*}) + (w_i C_i^{\sigma^*} + w_{i+1} C_{i+1}^{\sigma^*})$$
$$> w_i C_i^{\sigma^*} + w_{i+1} C_{i+1}^{\sigma^*}.$$

Thus $\sum_{i=1}^n w_i C_i^\sigma > \sum_{i=1}^n w_i C_i^{\sigma^*}$, which contradicts the fact that σ is optimal. □

So when we number the jobs in order of non-increasing weights we know that there exists an optimal solution satisfying this order.

4.2   Combinatorial Algorithm

We have split the problem up in two parts: we need to find a good sequence for the order on the job completions (secuencing problem) and then decide how to make the groups (partitioning problem). In Sect. 4.1 we solved the sequencing problem for some special cases. To find an optimal solution it remains to solve the partitioning problem. In Sect. 2 we showed that we can rewrite the problem as an LP to find the exact completion times, and hence make an optimal solution. In this section we develop a combinatorial algorithm that solves the partitioning problem in linear time. We first give a intuitive explanation of this algorithm, and thereafter a formal definition of the algorithm in pseudo code is given.

The algorithm finds an sequence of groups that fits in the block structure of an optimal solution for the given order. This sequence indicates which groups of jobs should be processed in what order. Again we assume w.l.o.g. that the order on the completion times is $C_1 \leq \cdots \leq C_n$. The algorithm determines for each job i, whether it is better to schedule this job in the same group as job $i - 1$ or to start a new group for job i. Hereto, we determine what the effect of job i is on the total weighted completion time is when it is scheduled in the same group as job $i - 1$ and also for the case when a new group is started for job i. When job i is scheduled in the same group as job $i - 1$ the contribution is $(\sum_{j \in G_k} w_j + \sum_{j=i}^n w_j)(p_i/s_{G_k})$, as it delays all jobs processed in or after group G_k. Here we use s_{G_k} to refer to the speed at which G_k is processed. When job i starts a new group the contribution is $(\sum_{j=i}^n w_j)(p_i/s_i)$, as it delays job i and all jobs that complete after job i. Therefore we determine whether or not the following equation holds:

$$\frac{\sum_{j \in G_k} w_j + \sum_{j=i}^n w_j}{s_{G_k}} \leq \frac{\sum_{j=i}^n w_j}{s_i}. \tag{1}$$

If so, then job i is scheduled to be processed together with the group of $i - 1$. Otherwise a new group is started. The pseudo code of this greedy algorithm can be found in Algorithm 1.

Theorem 4. *Algorithm 1 finds an optimal solution for a given order of job completions in JDMS.*

Proof. Suppose σ is an optimal schedule for an instance of JDMS with n jobs satisfying $C_1 \leq \cdots \leq C_n$. Let σ^* be the solution according to the algorithm. Then we want to show that a schedule according to the algorithm will give a solution with an equal objective value. Let job i be the first job in σ that is not scheduled according to the algorithm. Then there are two possible situations:

1. In σ, job i is scheduled in a new group G_k, whereas in σ^* it is still processed in group G_{k-1}.

Input: n jobs with processing requirements $p_1, \ldots, p_n$, weights $w_1, \ldots w_n$, speeds $s_1, \ldots, s_n$ and an order on the completion times $C_1 \leq \cdots \leq C_n$.
Output: an optimal sequence of groups $[G_1, \ldots, G_k]$
initialization: $G_1 = \{1\}, k = 1, E = w_1, s = s_1$;
for $i = 2$ *until* $i = n$ **do**

> **if** $(E + \sum_{j=i}^{n} w_j)s_i \leq (\sum_{j=i}^{n} w_j)s$ **then**
>> $E \rightarrow E + w_i,$;
>> $G_k \rightarrow G_k \cup \{i\}$;
>
> **else**
>> $E \rightarrow w_i$;
>> $s \rightarrow s_i$;
>> $G_{k+1} \rightarrow \{i\}$;
>> $k \rightarrow k+1$;
>
> **end**

end
Return $[G_1, \ldots, G_k]$

Algorithm 1. GREEDY ALGORITHM

2. Job i is scheduled in the previous group G_k, while it should be scheduled in a new group.

Suppose we are in the first situation: in σ, job i is scheduled in a new group G_k, whereas in σ^* it is still processed in group G_{k-1}. Then we change σ to σ' by merging G_k and G_{k-1}. Then in σ' , jobs $1, \ldots, i$ are scheduled the same as in σ^*.

We look at the total value that groups G_k, G_{k-1} and $G_k \cup G_{k-1}$ contribute to the objective value of σ and σ'. Here the contribution of a group S is not $\sum_{j \in S} w_j C_j$, but instead the time it takes to process all jobs in the group multiplied by the weight of all unfinished jobs at that point. As the other groups in σ' are the same as in σ, the contribution of these groups are the same in both σ and σ'. Let c be the function that computes the value that a group contributes to the objective value. Let W be the total weight of all the jobs scheduled after G_k. Then:

$$c_\sigma(G_k) + c_\sigma(G_{k-1}) = \left(\sum_{j \in G_{k-1}} w_j + \sum_{j \in G_k} w_j + W \right) \frac{\sum_{j \in G_{k-1}} p_j}{s_{G_{k-1}}}$$

$$+ \left(\sum_{j \in G_k} w_j + W \right) \frac{\sum_{j \in G_k} p_j}{s_{G_k}}, \quad (2)$$

$$c_{\sigma'}(G_k \cup G_{k-1}) = \left(\sum_{j \in G_{k-1}} w_j + \sum_{j \in G_k} w_j + W \right) \frac{\sum_{j \in G_k \cup G_{k-1}} p_j}{s_{G_{k-1}}}. \quad (3)$$

According to Algorithm 1 it should hold that:

$$\frac{\sum_{j \in G_{k-1}} w_j + \sum_{j=i}^{n} w_j}{\sum_{j=i}^{n} w_j} \leq \frac{s_{G_{k-1}}}{s_i}. \tag{4}$$

Combining the fact that $\sum_{j \in G_k} w_j + W = \sum_{j=i}^{n} w_j$ and (4) we know that:

$$\frac{\sum_{j \in G_{k-1}} w_j + \sum_{j \in G_k} w_j + W}{\sum_{j \in G_k} w_j + W} \leq \frac{s_{G_{k-1}}}{s_i}. \tag{5}$$

We rewrite (5), and as job i is the first job of G_k we can replace s_i by s_{G_k}:

$$\frac{\sum_{j \in G_{k-1}} w_j + \sum_{j \in G_k} w_j + W}{s_{G_{k-1}}} \leq \frac{\sum_{j \in G_k} w_j + W}{s_{G_k}}. \tag{6}$$

Multiplying both sides with $\sum_{j \in G_k} p_j$:

$$\left(\sum_{j \in G_{k-1}} w_j + \sum_{j \in G_k} w_j + W \right) \frac{\sum_{j \in G_k} p_j}{s_{G_{k-1}}} \leq \left(\sum_{j \in G_k} w_j + W \right) \frac{\sum_{j \in G_k} p_j}{s_{G_k}}. \tag{7}$$

Adding $\left(\sum_{j \in G_{k-1}} w_j + \sum_{j \in G_k} w_j + W \right) \frac{\sum_{j \in G_{k-1}} p_j}{s_{G_{k-1}}}$ yields:

$$\left(\sum_{j \in G_{k-1}} w_j + \sum_{j \in G_k} w_j + W \right) \frac{\sum_{j \in G_k \cup G_{k-1}} p_j}{s_{G_{k-1}}} \leq \left(\sum_{j \in G_k} w_j + W \right) \frac{\sum_{j \in G_k} p_j}{s_{G_k}}$$

$$+ \left(\sum_{j \in G_{k-1}} w_j + \sum_{j \in G_k} w_j + W \right) \frac{\sum_{j \in G_{k-1}} p_j}{s_{G_{k-1}}}. \tag{8}$$

Combining (2), (3) and (8) yields:

$$c_{\sigma'}(G_k \cup G_{k-1}) \leq c_\sigma(G_k) + c_\sigma(G_{k-1}). \tag{9}$$

Thus the value of our changed schedule σ' is smaller or equal than the objective value of σ, thus σ' is optimal as well. The proof for the second situation is similar and will therefore not be fully written out here.   □

5   Concluding Remarks

In this paper, we considered the problem JDMS in which jobs need to be scheduled preemptively on a single machine of which the speed varies with the number of jobs that have been completed. We showed this problem to be NP-hard and that the main issue is to decide in which order the jobs need to complete. Once this order is known, we can find the optimal schedule in polynomial time through a greedy algorithm. This algorithm uses a structural property that tells that the jobs are processed in blocks. For two special cases, JDMS with unit weights or unit processing times, we found the orders that guarantee us to find an optimal solution. The optimal order to complete unit weight jobs is shortest processing time and the one for unit processing times is largest weight.

One question that remains is how well the general problem can be approximated. In case that the speed is constant, it is well known that the WSPT rule that processes the jobs in order of non-increasing ratio of weight over processing time is optimal [15]. However, the following example shows that this order can be arbitrarily bad for JDMS.

Example 1. In this example, there are two jobs with $w_1 = 0$, $p_1 = \epsilon$ and $w_2 = p_2 = A$. According to the WSPT order job 2 precedes job 1, and the optimal schedule has value A^2. If we consider the opposite order the optimal schedule has value $\frac{A}{1+\epsilon}$. Letting ϵ go to 0 and A be arbitrarily large, we see that this ratio can be arbitrarily large.

Acknowledgements. We thank Urtzi Ayesta for helpful discussion after posing this open question during the Daghstuhl Seminar 13111 "Scheduling" in 2013. Furthermore, we thank the organizers of this seminar and Schloss Daghstuhl for providing the right atmosphere to facilitate research.

References

1. Albers, S.: Review articles: energy-efficient algorithms. Commun. ACM **53**(5), 86–96 (2010)
2. Alidaee, B., Ahmadian, A.: Scheduling on a single processor with variable speed. Inf. Process. Lett. **60**, 189–193 (1996)
3. Ayesta, U.: Scheduling (dagstuhl seminar 13111): scheduling with time-varying capacities. Dagstuhl Rep. **3**(3), 29 (2013)
4. Bekker, R., Boxma, O.J.: An M/G/1 queue with adaptable service speed. Stochastic Models **23**(3), 373–396 (2007)
5. Bertrand, J.W.M., Ooijen, H.P.G.: Workload based order release and productivity : a missing link. Prod. Plan. Control : The Manage. Oper. **13**(7), 665–678 (2002)
6. Ewalid, A., Mintra, D.: Analysis and design of rate-based congestion control of high-speed networks, i: stochastic fluid models, access regulation. Queueing Syst. **9**, 29–64 (1991)
7. Ewalid, A., Mintra, D.: Statistical multiplexing with loss priorities in rate-based congestion control of high-speed networks. IEEE Trans. Commun. **42**, 2989–3002 (1994)
8. Demers, A.J., Yao, F.F., Shenker, S.: A scheduling model for reduced CPU energy. In: FOCS, pp. 374–382 (1995)
9. Garey, M.R., Johnson, D.S.: Computers and Intractability: A Guide to the Theory of NP-Completeness. W.H. Freeman and Co., New York (1979)
10. Gawiejnowicz, S.: A note on scheduling on a single processor with speed dependent on a number of executed jobs. Inf. Process. Lett. **57**, 297–300 (1996)
11. Mandjes, M., Mintra, D.: A simple model of network access: feedback adaptation of rates and admission control. In: Proceedings of Infocom, pp. 3–12 (2002)

12. Megow, N., Verschae, J.: Dual techniques for scheduling on a machine with varying speed. In: Fomin, F.V., Freivalds, R.U., Kwiatkowska, M., Peleg, D. (eds.) ICALP 2013, Part I. LNCS, vol. 7965, pp. 745–756. Springer, Heidelberg (2013)
13. Boxma, O.J., Bekker, R., Borst, S.C., Kelly, O.: Queues with workload-dependent arrival and service rates. Queueing Syst. **46**(3–4), 537–556 (2004)
14. Ramanan, K.A., Weiss, A.: Sharing bandwidth in ATM. In: Proceedings of the Allerton Conference, pp. 732–740 (1997)
15. Smith, W.E.: Various optimizers for single-stage production. Naval Res. Logist. Q. **3**, 59–66 (1956)

Author Index

Printed in the United States
By Bookmasters

Printed in the United States
By Bookmasters